Center for Basque Studies
Current Research Series, No. 5

Innovation: Economic, Social, and Cultural Aspects

Edited by

Mikel Gómez Uranga and Juan Carlos Miguel de Bustos

Current Research Series, No. 5

Center for Basque Studies
University of Nevada, Reno

Published in conjunction with the University of the Basque Country
UPV/EHU

Center for Basque Studies
UNIVERSITY OF NEVADA, RENO

eman ta zabal zazu
Universidad del País Vasco
Euskal Herriko Unibertsitatea

Current Research
Selections of the ongoing work done by the faculty of the University of the Basque Country (UPV/EHU), www.ehu.es/

Current Research Series, No. 5
Center for Basque Studies
University of Nevada, Reno
Reno, Nevada 89557
http://basque.unr.edu

Cover design based on engravings by Eduardo Chillida and Jorge Oteiza.

Library of Congress Cataloging-in-Publication Data

Innovation : economic, social, and cultural aspects / edited by Mikel Gomez Uranga and Juan Carlos Miguel de Bustos.
p. cm. -- (Current research series ; No. 4)
"Published in conjunction with the University of the Basque Country."
Includes bibliographical references and index.
Summary: "Collection of articles on economic, social, and cultural aspects of innovation by researchers at the University of the Basque Country"--Provided by publisher.
ISBN 978-1-935709-12-1 (pbk. : alk. paper)
1. Creative ability in business. 2. Diffusion of innovations. I. G?mez Uranga, M. (Mikel), 1948- II. Miguel de Bustos, Juan Carlos.

HD53.I563 2011
338'.064--dc23

2011037165

Contents

Introduction

Innovation, Social Impacts, and Creative Spaces

Mikel Gómez Uranga and Juan Carlos Miguel de Bustos

Translated by Jennifer R. Ottman

The most used and most widespread concept of innovation has its gravitational center in business, in such a way that the entire constellation made up of innovation agents and activities moves under its influence. Businesses use or create technical resources and procedures and are organized in different ways, both internally and externally. From the business system come products, which have to be prepared for acceptance by the market. In all phases, procedures, and activities, innovations appear, which are defined in OECD (Organisation for Economic Co-Operation and Development) manuals and accepted as standards by the "economic world" as well as by all governments. This customary vision of innovation corresponds to or is inspired by the writings of one of the most prolific authors in the field of economics, Joseph Schumpeter. Modern conceptions of innovation hark back to Schumpeter, specifically to his book *Business Cycles: A Theoretical, Historical, and Statistical Analysis of the Capitalist Process* (1939). In this work, economic growth (over the long term) is explained through the appearance of innovations. For this celebrated author, innovation belongs to the field of economic (business) decisions and is distinguished from *invention*, which because it belongs to the field of technology, remains outside the economic sphere. Schumpeter's inheritance is found in two prominent places:

- In the new evolutionary paradigms of the economy of innovation and technical change that began to spread in the 1980s. There is agreement among authors belonging to this current that the spread of innovations demands social, organizational, and institutional change (for example, Chris Freeman, Richard R. Nelson and Sidney G. Winter, Giovanni Dosi, Nathan Rosenberg, Paul A. David, David C. Mowery, and Bengt-Åke Lundvall). They also seek to explain how these changes coevolve with other changes in technology.

- And in the significance assigned to the term *innovation* in economic, business, political, and administrative spheres all over the planet. In this regard, the OECD is a body that establishes, as a standard, the exact meaning that should be given to the term *innovation* (OECD 2005, 46).

The earlier vision of innovation, which we can label *economistic*, has had to be complemented in a variety of ways, among which we will initially highlight two: understanding the context and the conditions in which innovation is generated, giving rise to the concept of an *innovation system;* and understanding that scientific knowledge is part of the business system, both as a supplier and by way of its procedural methodology. Hence, the sphere of innovation escapes that of the business system and expands considerably.

In business logic, but also outside that logic, there exist activities related to innovation that do not have a direct economic result. Along these lines might be found, for example:

- Part of what has been called *non-business research* and other activities necessary for innovation, as indicated in the OECD's *Frascati Manual.*
- In the synthetic indicator used by the European Commission to measure innovation, the SII (Summary Innovation Index), factors such as educational level and Internet use among the population are considered.

It is possible to consider innovations found outside the business orbit and even on the margins of the market sphere, among which it is appropriate to mention those innovations generated in nonprofit organizations, social networks, the public sector, NGOs, and so on.

Definitions of innovation different from those inspired by Schumpeter and market logic might be considered, such as defining innovation as new forms, as new procedures or ways of acting, or of organizations that contribute to obtaining social or other kinds of results (not necessarily economic ones).

In this book, we analyze *creative spaces* by way of the two words that make up the concept:

- We consider *space* from two perspectives: as the location of an activity and as a territory in which innovation activities take place.
- We associate *creative* with the concept of creativity, but here we confront the term's complexity and scope. For this reason, we will need to engage in additional explanation.

Since the eighteenth century and the Romantic era, the concept of *creativity* has found its most genuine expression in literary creation and in a select group of various arts such as music, architecture, painting, and so on. In recent decades, in some sectors, such as research on design, advertising, fashion, artificial intelligence, and the cultural industries, the concepts of innovation and creativity have been united. In these cases, creativity is considered as a source of competitiveness and is translated into economic results. Innovation, on the other hand, is linked more closely to human activity, as is increasingly demonstrated by the interest aroused by the term *creative industries*.

The concept of creative industries, which arose in Great Britain in the second half of the 1990s as a tool of industrial policy, has begun to be used worldwide, in all spheres: administration, international organizations, and elsewhere. According to this approximation, creative industries are considered to be—in addition to cultural industries (the press, radio, television, books, films, music, and adevertsing)—industries such as design, art, decorative arts, fashion, antiques, and even software. From various perspectives, creativity is associated with innovation, creation being seen as a necessary prerequisite for innovating.

Creativity refers to the act of producing new ideas, ways of thinking, and/or ways of approaching actions or activities. On the other hand, innovation requires, in addition to the generation of new ideas (the creative act), the application of those ideas in a specific context.

In organizations, the term *innovation* refers to the process by which an organization generates new creative ideas and turns them into something novel, useful, and viable, through commercial products, services, and business practices. In each case, creation would be considered an element prior to the process of innovation. But innovation and creativity, although they may be the fruit of individual realizations, are always anchored in a particular society, where flows of ideas are established among various people, organizations, and institutions (Pratt 2008).

Innovation activities cannot fail to include those that entail significant changes in the sphere of consumption via the Internet, as well as the use of unprecedented new forms of advertising, or new visions and uses of technology in art, and so on.

Contributions

In "Social Capital and Innovation Relationship: Two Empirical Tests," authors Jose M. Barrutia and Carmen Echebarria seek to establish an empirical relationship between social capital and innovative activity in the regional arena. The authors measure the impact of social capital on innovation. The measurement of social capital (from a double economic and sociological perspective) is very infrequent and not exempt from problems.

In order to construct their indicators, the authors explore different dimensions of social capital. We were pleasantly surprised by their ability to successfully capture in quantitative indicators certain "values" found on the level of human relationships, such as trust and tie strength.

The authors conclude that there is a positive correlation between social capital and innovation, but they also maintain in their article that intense social relationships or ties may have negative effects on innovative capability because they may demand costly time investments from their subjects. This "nuanced" solution reached by the authors is highly original in light of the current literature on innovation. It offers guidance to those making decisions about regional innovation policies on the need to act in some cases to increase social capital, and in others to decrease the costs associated with a given level of relations.

In "Social Innovation, Locational Capital, and Informality in Latin American Cities" authors Pedro Abramo and Arantxa Rodríguez discuss the evolution of the "informal city" in various places in Latin America. This process of the growth of cities for poor populations is characterized by the authors as a social innovation. Urban territory is the result of social innovations that are reinforced and accumulate over time, but these processes do not result in formal or legal institutionalization with the passage of time.

Informal territory persists and coexists with the "formal city," and a monetarized urban real estate market comes into being outside the legal structures, stemming from a prior social innovation.

It seems to us that the authors make an interesting contribution from the perspective of institutional economy and that their analysis can shed more light on urban and redistributive policy challenges potentially faced by the authorities in many cities around the world.

In "The Precautionary Principle: Regulation Needed to Confront the Risks Associated with Nanotechnology," Alejandro Arizkun Cela and Iñaki Lasagabaster Herrarte limit and specify the meaning of the "precautionary principle." This principle is intimately associated with the uncertainties that accompany the development of science and technology. The application of this principle has enormous significance; in fact, we human beings are staking nothing more or less than the salvation of the planet on it. The novelty of this article is the application of this principle to the development of nanotechnology. The authors introduce this cutting-edge sector in order to sketch the potential problems derived from nanomaterials' toxicity, something that could have a negative impact on human health and the environment in view of the consumption of products derived from nanotechnology. In conclusion, the authors note that nanotechnology is not subject to specific legal regulation, in the European case in particular.

In "New Technologies in the Globalized Economy: Patents as Tools for Measuring and Comparing Innovation Systems," authors Itziar Urizar and Virginia Fernández, taking regional and national innovation systems as their methodological foundation, empirically analyze those systems that correspond most closely to the paradigm of the globalization of today's economies, in particular those of the so-called emerging countries (such as China and India). The authors present in their study a useful and carefully worked out report, using the most relevant indicators for obtaining familiarity with any innovative system or organization, such as research and development (R&D) expenditure and patents.

The authors' analysis consists of comparing different countries through the use of customary international indicators. Among the results demonstrated in the article, software production and research is highlighted in the case of India, as is Asia's more rapid advance in the chief indicators, surpassing the United States and far ahead of Europe. Finally, the authors highlight the central aspects of the progress of the Chinese system of science and technology.

Services are an increasingly large part of industrial activity, and for this reason, international organizations have paid greater attention to innovation in services in recent years. In the chapter by Mikel Zurbano and Goio Etxebarria, "Innovation in Services: An Application in the Case of KIBS," the authors observe that this phenomenon historically has been considered in discussions of innovation almost exclusively in its technological aspects.

The innovation in services that is considered here has to do chiefly with business activities circumscribed to actions in and affecting the market, such as in the sectors of telecommunications, financial services, advertising, consulting, and so on. The authors choose the case of "knowledge-intensive business services" for their study. In this sector, the standard profile is represented by small- and medium-sized firms that act as suppliers to others, especially in the software area. In some cases, such firms are established by outsourcing services that were previously integrated into large firms' R&D departments.

In "Corporate Strategies, Competition, and Technological Innovation among the Operators of Maritime Transport Services," authors Iñaki Alcedo Momoitio and José Moreno Diaz study innovation in a sector of crucial importance for a globalized economic system. This sector has been in some ways neglected, despite its importance, by academic publications.

Changes in the maritime sector are also associated with the rapid development of trade to and from emerging countries. The international financial crisis has pushed the sector to implement significant changes, and especially to innovate. Innovation is at work in various areas of the maritime container business, such as the following:

- Organizational changes in the management of firms as well as in the logistical sector.
- In the area of vendor/client relations (innovation in marketing techniques).
- Changes in economic organization (of Schumpeterian inspiration) such as mergers, acquisitions, and new alliances.

We may say that this chapter fits very well in the book as a whole because the study of this sector brings together in a single framework an exhaustive variety of innovation profiles, which could serve as a case study for any general discussion of innovation.

Higher-education institutions also find themselves immersed in change. These changes stem not only from the use of technological tools, but also from organizational changes. In "Innovative Management and Changes in Organizational Culture at the University of the Basque Country," Francisco Panera Mendieta and Sabino Ayestarán Etxeberria offer a case study of the University of the Basque Country, describing the stages the university has passed through on its path toward management innovation, both in the sphere of training for quality and in certification (ISO and EFQM standards), in which it has won extensive recognition for quality through such certifications and awards.

This is especially important in universities: external processes are increasingly important, yet the culture has been characterized as clannish and egocentric. The issue is complicated at the University of the Basque Country with three campuses in three different provinces, each with its own very different social and historical development. Change is complicated and difficult. In this case, it is a matter of changing the organizational culture to one that is more hierarchical and that has internalized market aspects as a fundamental requirement for improving the university's own management, as well as relations with its external environment. The central driver of change in modeling a culture of quality and quality management is training of a special kind, *action training*, which enables the immediate application of what has been learned.

A significant turning point came in 2001 because it inaugurated a new stage, stretching over several years, in which the number of centers involved gradually increased, all working in relation to the model put forward by the EFQM (European Foundation for Quality Management). The result was that year by year, specific training centers were recognized or received awards for their management.

In "The Technological 'Interface' in Contemporary Art," Josu Rekalde analyzes and contextualizes Basque art in the age of networks and electronics. It is not a matter only of changes or improvements in the tools used by the artist but also of the fact that technology is transforming the relationship between those who create and those who provide the audience for the artistic act. The concept of spatiality is expanded by the possibility of creating and re-creating unimaginable spaces, even as the result of algorithms. In addition, the temporal axis can be explicitly included. In this way, the relationship between creators is established in a form that is not just experience, but a process in which the individual's experience is situated.

New information and communications technologies enable the globalization of artistic experiences and pose a central question, the relationship between the local artist, who endows his or her work with meaning, and the global spectator. The use of electronic and networked means of expression should not cause us to forget that artists are individuals, just like those who go to see and experience art, so that although the artistic act is being transformed, it is fundamental not to forget its human aspect.

The author profiles the history of Basque "technological" art and locates its beginnings in Iruñea-Pamplona in 1972 with the support of patrons from other creative spheres, such as music, even if it acquired firmer foundations at the beginning of the 1980s with the institutionalization of video festivals, in which the author was himself a participant.

The arrival of the Internet has entailed significant changes in the field of journalism. Experiments are being made with new kinds of content, new narratives, new business models. It is also an opportunity to attract new audiences, such as young people, who consume little of the print press. Ainara Larrondo Ureta, the author of "Innovation and Creative Content in Internet Media," distinguishes several stages. First was that of *shovelware*, in which the content of online editions was based on already published materials that were recycled and adapted for web distribution. Subsequently, dedicated editorial teams were set up, generally composed of young professionals, who began to experiment and develop new journalistic formats.

What has been called *convergence* entails online innovation through the use of tools of all kinds (video, hypermedia, and so on). Content is affected by innovation in the form of infographics, the coverage of mega-events, special reports, and so on.

These changes affect the way of creating content (production routines), the way of presenting it, the coverage of events in different ways, and so on, prefiguring changes in the audiovisual narrative itself. Even the roles of journalist and editor may change if we think about the fact that opinions about content are generated in parallel, as in Eva Ferreras's chapter on Menéame.

The Internet has become a fundamental tool in election campaigns, especially since Obama's campaign, in which social networks began to be used on a massive scale.

It is reasonable to think that the Internet entails a change comparable to that of the birth of television. In this case, social networks—YouTube,

Twitter, Facebook, Flickr, and Myspace—make it possible to reach younger target audiences, who are often those least likely to vote. The fundamental tool, in addition to email, is social networking. "Social Networks in the British Elections: Minimalist Efficiency," by Iker Merodio and Estefanía Jiménez, analyzes the British elections of April and May 2010 (in particular, the choice of communicative channels), as well as the use of sites on which the three candidates and their parties were presented.

This analysis makes it possible to study the value placed on social networks by each candidate's team, as well as the differences in their use and reception—that is, the number of users.

Of the three candidates, the most novel option was that of the Liberal Democrats, who initially were facing the most negative forecasts and who obtained the most concrete results, revolutionizing the campaign, both within and beyond the social networks. Their leader's dominance on television led the other media and communicative channels (also the Internet) to increase their attention to Nick Clegg. He was by a clear margin the candidate who made the greatest effort to have an Internet presence in general and a social-networking presence in particular, without neglecting more traditional platforms.

Web 2.0, or collaborative networking, entails great changes because it enables sharing and collaboration in different spheres and situations. "New Proposals for the Creation of an Informative Agenda: The Case of Menéame," by Eva Ferreras, analyzes one of the chief tools used in Spain for ranking news.

Menéame, the subject of her case study, was born in 2005, inspired by Digg, and defines itself as a social news promoter. It is a collaborative website based on the social ranking of content already published elsewhere, in online newspapers, blogs, and so on.

The interest of these sites is that they become a kind of editor of the existing editors, adding value by putting a value on what those editors have already published. They become a complement to what has already been edited, a complement of a kind impossible outside the Internet because it requires speed.

The site is not only a tool for users, but also offers information to editors, who can establish guidelines and protocols conducive to making their news items more highly rated than those of the competition. At the same time, there is a danger that in this quest to position one's own items at the top of the rankings, the task of writing and editing becomes perverted by the search for the keys to the top spots.

References

OECD (Organisation for Economic Co-operation and Development). 2005. *Oslo Manual: Guidelines for Collecting and Interpreting Innovation Data*. 3rd ed. Paris: OECD/Eurostat.

Pratt, Andy C. 2008. Innovation and Creativity. In *The SAGE Companion to the City*, edited by T. Hall, P. Hubbard, and J. R. Short. London: Sage. At LSE eprints (2009), http://eprints.lse.ac.uk/20718/1/Innovation_and_creativity_%28LSERO%29.pdf.

Schumpeter, Joseph A. 1939. *Business Cycles: A Theoretical, Historical, and Statistical Analysis of the Capitalist Process*. New York and London: McGraw Hill Book Co.

1

Social Capital and Innovation Relationship: Two Empirical Tests

JOSE M. BARRUTIA and CARMEN ECHEBARRIA

The importance of social capital as an antecedent of innovation has received much theoretical attention during recent years (Landry, Amara, and Lamari 2002). In particular, social network theories of innovation hypothesize a positive and direct relationship between social capital and innovation. Nevertheless, empirical evidence regarding the social capital–innovation relationship is scarce in a regional context. Some interesting exceptions are the contributions of Richard Florida, Robert Cushing, and Gary Gates (2002); Sjoerd Beugelsdijk and Ton van Schaik (2005a, 2005b); Lee Fleming and Matt Marx (2006); and Christoph Hauser, Gottfried Tappeiner, and Janette Walde (2007), who generally find a positive relationship between weak social ties and innovative activity.

A main reason for empirical effort being very limited in this area is the lack of agreement regarding the content of the concept of social capital and the appropriate way of measuring it (Beugelsdijk and van Schaik 2005a). Social capital is a complex construct that (as was proposed by Nahapiet and Ghoshal 1998) has a structural, a relational, and a cognitive dimension (and various subconstructs might also exist inside these dimensions). Consensus exists regarding the multidimensionality of the concept, but concrete proposals to measure social capital from a multidimensional perspective are almost nonexistent (Barrutia and Echebarria 2010).

When research overcomes the previous limitation by using mainly individual indicators of social capital, regional researchers have focused on the social capital–gross domestic product (GDP) direct relationship instead of the more manageable social capital–innovation relationship. This is the case, for example, of Stephen Knack and Philip Keefer (1997); Robert D. Putnam (1995); Paul J. Zak and Stephen Knack (2001); Sjoerd Beugelsdijk and Sjak Smulders (2003); and Sjoerd Beugelsdijk and Ton van Schaik (2005a, 2005b). This stream of research has made an important contribution in relation to the impact of the various dimensions of social capital on innovation. In particular, a great emphasis has been placed on studying and discussing the diverse impact of bridging and bonding on regional growth. On the contrary, empirical research to contrast the social network theories of innovation at a regional level is practically nonexistent (for exceptions, see Hauser, Tappeiner, and Walde 2007; Barrutia and Echebarria 2010). To cover this research gap, more emphasis is needed on empirically studying the relationship between social capital and innovation, which is the main aim of this chapter.

Empirical studies of other units of analysis (such as individuals, teams, businesses, organizations, small world, and cities and regions) have mostly shown that some proxies of the various dimensions of social capital have a positive (and usually linear) effect on innovation, as has been conceptually suggested, for example, by Philip Cooke, Nick Clifton, and Mercedes Oleaga (2005); Mourad Dakhli and Dirk De Clercq (2004); and Réjean Landry, Nabil Amara, and Moktar Lamari (2002). Yet, some recent research has found a nonsignificant or even negative relationship (Ahuja 2000). The research has also suggested that social capital has its limits, and the social capital–innovation relationship might have a diminishing return shape (for example, Mcfadyen and Cannella Jr. 2004) or an inverted U-shape (for example, Uzzi and Spiro 2005; Leenders, van Engelen, and Kratzer 2003). The main idea behind these findings is that creating new relationships is costly, and maintaining existing ties also consumes time, energy, and financial resources. As tie strength is increased, there is less time left to seek out new resources that might lead to good ideas to fuel innovation (Zheng 2010). This line of reasoning is consistent with the thinking of regional researchers who have emphasized the relevance of weak ties for growth, such as Gerald Schneider, Thomas Plümper, and Steffen Baumann (2000); Sjoerd Beugelsdijk and Sjak Smulders (2003); Jose M. Barrutia and Carmen Echebarria (2007b); and Richard Florida (2003), and have shown the limits of trust and strong ties.

Building on the social network theory of innovation, this study seeks to contribute insights to this stream of research by empirically testing the impact of social capital on innovation in the context of the NUTS1 and NUTS2 (Nomenclature of Units for Territorial Statistics) European regions. Our results support the existence of a significant direct and positive effect of social capital on patents (as a measure of the innovation outcomes) up to a tipping point. The social capital–innovation relationship therefore shows an inverted U-shape. Although this finding might be considered relatively surprising, it is consistent with the previously mentioned recent empirical research carried out mainly on nongeographical levels of analysis. It suggests that some subconstructs of social capital, such as tie strength, involve maintenance and opportunity costs, and that social capital has positive effects but also has limits. Human relations are time-consuming, and time is not an unlimited resource. If the priority is to build social capital, other targets might be overlooked.

Although our research question focuses on social capital, we build on social network theories of innovation and thus consider social capital as a moderator of the relationship between knowledge/learning effort-related metrics (R&D expenditure and intellectual capital) and innovation outcomes. The first objective of our research is to study the shape of the social capital–innovation relationship in the European regions.

Concepts and Metrics

The main objective of this research is to empirically validate social network theories of innovation in the context of the European regions by focusing on social capital and R&D effort as the main drivers of innovation. Table 1.1 includes summarized descriptions of the variables used in this research.

We propose the GDP on R&D in Purchasing Power Standards (PPS) to proxy learning effort and absorptive capacity. The underlying concept, which is sometimes misunderstood, "comprises creative work undertaken on a systematic basis in order to increase the stock of knowledge, including knowledge of man [sic], culture and society and the use of this stock of knowledge to devise new applications" (OECD 2002, 63). This is a simple and robust metric, which measures real effort in terms of actual cost expended for the generation of new marketable knowledge or learning. Furthermore, it is a structural metric that moves slowly and with a generally upward trend within the European context, so it is possible to proxy the knowledge that has accumulated within a particular geographical area and its absorptive capacity. We use Eurostat data.

There is a variety of interrelated definitions of social capital (see Adler and Kwon 2002; Borgatti and Foster 2003; and Zheng 2010 for differently focused recent works). A comprehensive social capital framework was proposed by Janine Nahapiet and Sumantra Ghoshal (1998). Their framework integrated various components of social capital into three dimensions: the structural dimension, the relational dimension, and the cognitive dimension. The *structural dimension* refers to configurations and patterns of connections between people, addressing properties such as network density, connectivity, and hierarchy. Social capital is a recent yet significant addition to the list of innovation-inducing factors (Zheng 2010).

Social network theories also echo that the locus of innovation is no longer the individual or the firm, but the network in which a firm is embedded, which shifts attention to social capital (Powell, Koput, and Smith-Doerr 1996). Knowledge is viewed as an intrinsic part of the surrounding social, political, and economic context, and knowledge creation occurs through the integration of resources (Collinson 2000). In these theories, knowledge is embodied in networks and communities, and social capital becomes an essential ingredient for understanding innovation. So a high level of social capital is viewed as having a positive effect on regional innovation and growth (Storper 1995; Barrutia, Aguado, and Echebarria 2007; Barrutia and Echebarria 2007a; Echebarria, Barrutia, and Aguado 2009). It has been argued that social capital prompts the efficiency and effectiveness of the innovation process. According to Peter Maskell (2001), social capital helps reduce malfeasance, induces the volunteering of reliable information, causes agreements to be honored, enables employees to share tacit information, and places negotiators on the same wavelength. Efficiency, then, is mainly improved by reducing transaction costs (between firms, and between firms and other actors) and the costs of managing the innovation process. And greater effectiveness is achieved as more (suitable and reliable) quantity and quality of knowledge is accessed. Knowledge can more easily be transferred and utilized within a community made up of firms that understand the same language and share norms and codes (Maskell 2001). On the contrary, a low level of social capital implies lack of coordination, duplications of effort, and costly contractual dispute (Fountain and Atkinson 1998).

There is no consensus on how to measure social capital. Previous literature has concentrated on describing the concept of social capital and its implications. Following a sociologically driven approach and using data from the European Value Studies (EVS—see www.europeanvalues.nl), a regional measure of social capital has been proposed: the Beugelsdijk and van Schaik (2005a) social capital index, which uses the Eurostat NUTS1 definition of regions. These authors' index incorporates trust, and active and passive group membership (respondents who are members of a particular association but do not perform voluntary work for that body). We use this index in test 1.

From a more economic-driven approach, some authors (for example, Coleman 1990; Mcfadyen and Cannella Jr. 2004) refer to social capital as investment in social relations with expected returns in the marketplace. For the purposes of this research, we adopt the approach proposed by Francisco Pérez et al. (2006), incorporating methodology from previous work (Glaeser, Laibson, and Sacerdote 2002; OECD 2001). Investment in social capital is modeled around three basic hypotheses: (1) cooperation is favored by the economic incentives deriving from the expected increasing income resulting from continued growth; (2) individuals' incentives to cooperate are strengthened/weakened by two factors: a) the effective opportunities for participation in the results, and b) a culture of fulfillment of duty or of reciprocity that is increased and transmitted through access to education and the improvement of the population's human capital; and (3) the effects of cooperation are extended if the density of the trust relationship networks between individuals is high. This density of trust relations is favored by the smooth running of long-term economic relationships when uncertainty is inevitable and information costs, such as employment and financial relationships and markets for durable goods, are high.

All individuals decide their future investment path in social capital over T years, during which they expect to participate in the social network, maximizing the expected income from the investment. In equilibrium, each individual makes the marginal cost of investing in social capital equal the benefits expected from the investment (for a derivation, see Pérez et al. 2006). In equilibrium, the social capital is defined by the following:

$$KS_t = \frac{\beta}{\gamma} c(1 + \lambda(N-1)) \frac{(1-G)}{C'(Is_{it})} \frac{1-\left(\delta/(1-\rho)\right)^{T-t}}{1+\rho-\delta}$$

The variables considered and the proxies used for measuring social capital are as follows:

- *c:* Degree of connection of the social network (proxy: ratio Credit/GDP)
- *C'(Is):* Marginal cost of investment in social capital (proxy: 1—proportion of the population of working age with at least secondary education)
- *G:* Index of inequality in distribution of income (proxy: Gini Index)
- δ: Rate of survival of social capital stock (proxy: 1—rate of depreciation)
- ρ: Discount rate (proxy: constant discount rate: 4%)
- *T–t:* Number of expected years of participation in the social network (proxy: life expectancy for the average age of the population)
- *N:* Dimension of the social network (proxy: number of people in work)
- λ: Degree of reciprocity; β: Contribution of social capital to generation of income; γ: Contribution of work to generation of income (assumed to be constant, the social capital being calculated in relative terms)

We use this economic-driven metric in test 2, which considers Spanish and Italian NUTS2 regions. We use the Spanish data provided by the BBVA Foundation (2007) and compute the Italian data following the same methodology.

The outcome variable considered is the number of patent applications to the European Patent Office (EPO) per million inhabitants. Eurostat summarizes the main weaknesses that are usually attributed to this metric: (1) not all inventions are patented and not all patents have the same value; (2) the value distribution of patents is skewed because many patents have no industrial application; (3) inventions that are protected by other means or inventions, whose value does not warrant the costs of patenting, are not patented; (4) the propensity to patent differs across countries and industries; and (5) changes in patent law and regulations result in lower comparability over time. Despite these limitations the number of patent applications is the best metric available to measure innovation as has been suggested by Zvi Griliches (1995) and Manuel Trajtenberg (1990), among others. Specifically, the number of patents seems to be a good proxy for measuring the most relevant innovation outcomes because it includes the most radical and long-term profit-impacting innovations (i.e., innovations that are new to the firm and the world). Most investigations use the number of patents granted as the innovation outcome (see Shan, Walker, and Kogut 1994; and Ahuja 2000). It has been said that the propensity to patent depends on the sector. For instance, Anthony Arundel and Isabelle Kabla (1998) found that 80 percent of innovations were patented in the pharmaceutical sector compared to only 8 percent in the textile sector. Therefore, we include the presence of high-tech sectors in the region as a control variable.

Table 1.1. Variables

Concept	*Variable*	*Description*	*Source*
Patents (P)	Number of patent applications to the European Patent Office (EPO) per million inhabitants	Applications filed directly under the European Patent Convention or applications filed under the Patent Co-operation Treaty and designated to the EPO (Euro-PCT)	Data originate from the patent database PATSTAT hosted by the European Patent Office (EPO). Eurostat (mean 2000–2004).
Research/ Learning Effort (R&D)	Gross domestic expenditure on R&D in Purchasing Power Standards	Business enterprise expenditure, Higher education expenditure, Government expenditure, and Private nonprofit intramural expenditure on R&D	Data originate from the national R&D surveys, which are based on the *Frascati Manual*, OECD (2002). Eurostat.
Social Capital (SC)	Social Capital (Sociological Approach)	Social Capital Index	Beugelsdijk and van Schaik, 2005a. Data from European Values Study (1999/2000).
	Social Capital (Rational Choice Approach)	Mean of relations value for Spain = 100	BBVA Foundation (2007). For the Italian case: author's own work. Raw data have been obtained from the Italian Central Bank (Bank of Italy, Banca D'Italia).

Models To Be Tested and Empirical Results

Empirical models include the factors that may affect patents for region as explanatory variables: (1) R&D on PPS (R&Di); (2) Social Capital (SCi). Outcome variable is patents per million of inhabitants (Pi). Models to be tested are these:

1. Linear model:

$$P_i = \alpha + \beta_1 \, R\&D_i + \beta_2 \, SC_i + \varepsilon_i$$

2. Diminishing returns models:

$$P_i = \alpha + \beta_1 \, R\&D_i + \beta_2 \, (SC_i)^{1/2} + \varepsilon_i$$

$$P_i = \alpha + \beta_1 \, R\&D_i + \beta_2 \log (i) + \varepsilon_i$$

3. Inverted U-shape model:

$$P_i = \alpha + \beta_1 R\&D_i + \beta_2 \, SC_i + \beta_3 \, (SC_i)^2 + \varepsilon_i$$

Test 1 (NUTS 1, 54 European Regions, Sociologically Driven Metric)

Data were analyzed using Stata 10 statistical software. We checked for normality and linearity, and appreciated that the social capital–patents relationship seemed to be nonlinear. So it seemed evident that to test a diminishing return or an inverted U-shape relationship was appropriate.

After preliminary regressions were developed, the Breusch-Pagan/Cook-Weisberg test for heteroscedasticity showed lack of homocedasticity (chi2 (1) = 34.36, Prob.>chi2 = 0.0000). This diagnostic was confirmed after plotting the residuals versus the fitted (predicted) values. These problems were mainly related to the non-normality of the outcome variable, which was transformed by using a log function. Robust standard errors can deal with concerns about failure to meet regression assumptions, such as heteroscedasticity, or some observations that exhibit large residuals, leverage, or influence (Chen et al. 2003). Three different models were tested to contrast the social capital–innovation relationship: (1) an increasing linear model; (2) a diminishing returns model; and (3) an inverted U-shape model.

Table 1.2. Determinants of innovation outcomes

Specification	(1) Linear	(2) Dimin. return ($SC^{1/2}$)	(3) Dimin. return (Log)	(4) Inverted U-shape	(1) Linear	(2) Dimin. return ($SC^{1/2}$)	(3) Dimin. return (Log)	(4) Inverted U-shape
Variables	TEST 1: NUTS 1, n = 54, sociologically driven metric				TEST 2: NUTS 2, n = 35, economically driven metric			
Social capital (SC)	0.104 (0.167)			0.531** (0.010)	0.111** (0.478)			1.76*** (0.001)
SC^2				-.433*** (0.008)				-1.68*** (0.002)
$SC^{1/2}$ (or Log)		0.121* (0.070)	0.126* (0.062)			0.172 (0.255)	0.226** (0.109)	
R&D	0.602*** (0.000)	0.588*** (0.001)	0.589*** (0.001)	0.550*** (0.001)	0.381** (0.033)	0.353** (0.047)	0.330*** (0.062)	0.364** (0.049)
Constant	-41.35 (0.111)	-88.64** (0.022)	-88.64** (0.015)	-79.67** (0.013)	-11.88 (0.564)	-42.38 (0.260)	-139.98* (0.099)	-132.19** (0.020)
F-statistic	11.34***	12.89***	13.31***	8.42***	3.84**	4.13**	4.89**	5.59***
R-squared	0.4217	0.4259	0.4255	0.4416	0.1971	0.2106	0.2278	0.3511
Adj. R-sq.	0.3990	0.4033	0.4030	0.4081	0.1469	0.1613	0.1796	0.2883
Tipping point				58.07				130.10
Conf. Int.				(28.4, 87.7)				(105.8, 154.3)

*** significant at 1%; ** at 5%; *at 10%
Standardized coefficients (beta) are reported for variables.
p values are reported in brackets (robust std. errors).

Main results are reported in table 1.2 (left side). Specifications 1 to 3 compare the three models and show that the inverted U-shape model offers the best results. The inverted U-shape model and the social capital and R&D expenses variables are highly significant, and adjusted R-squared is the largest of the three models (0.4081). The tipping point is 58.07. The sociologically driven metric of social capital ranges from 0 to 100, although most countries present values of social capital below 50.

Test 2 (NUTS 2, 35 Spanish and Italian Regions, Economic-Driven metric)

After preliminary regressions, we tested the three models by using Stata 10 software. Results of test 2 are reported in table 1.2 (right side). Specifications 1 to 3 compare the three models and show that the inverted U-shape model offers the best results. The inverted U-shape model and the variables social capital and R&D expenses are highly significant, and adjusted R-squared is the largest of the three models (0.2883). The tipping point is 130.10. The values of the economic-driven metric of social capital range from 57 to 155.

Conclusions

The research question refers to the social capital–innovation outcomes relationship. A sociologically driven and an economic-driven metric are used to measure social capital and empirically contrast the impact of social capital on innovation outcomes. They show predictive validity to explain innovation outcomes. But research results also show that social capital has its limits. Specifically, the social capital–innovation relationship shows an inverted U-shape.

This finding might be considered relatively surprising, but it is consistent with recent empirical research carried out mainly at nongeographical levels of analysis, and suggests that some subconstructs of social capital, such as tie strength or trust, involve maintenance and opportunity costs and that social capital has positive effects, but also has its limits. The underlying idea is that human relations are time-consuming, and time is not an unlimited resource. If the priority is to build social capital, other targets should be neglected.

These conclusions are also implicit in the regional literature bridging-bonding debate. For instance, Richard Florida (2003) suggests that the "creative class" (people engaged in work whose function is to "create meaningful new forms") prefers weak ties. He concludes that social structures that historically embraced closeness may now appear restricting and invasive, and that where strong ties among people were once important, weak ties are now more effective. These remarks are consistent with the literature dealing with concepts of regional innovation systems, learning regions, and regional learning, which emphasize relatively weak (professional) ties as explanatory factors of successful innovation.

Finally, other suggestions for politicians emerge from this study. Investment effort in R&D is not in itself enough to achieve innovation outcomes in regions with low levels of social capital. R&D input should be accompanied by efforts to prompt social capital. Some Italian regions are good examples of the low impact of R&D investment input on innovation outcomes because of the almost or total absence of social capital. Investment rankings, which are very popular in political circles, should therefore be used with care.

References

Adler, Paul S., and Seok-Woo Kwon. 2002. Social capital: Prospects for a new concept. *Academy of Management Review* 27 (1): 17–40.

Ahuja, Gautam. 2000. Collaboration networks, structural holes, and innovation: A longitudinal study. *Administrative Science Quarterly* 45 (3): 425–455.

Arundel, Anthony, and Isabelle Kabla. 1998. What percentage of innovations are patented? Empirical estimates for European firms. *Research Policy* 27 (2): 127–141.

Barrutia, Jose M., Itziar Aguado, and Carmen Echebarria. 2007. Networking for Local Agenda 21 implementation: Learning from experiences with Udaltalde and Udalsarea in the Basque Autonomous Community. *Geoforum* 38 (1): 33–48.

Barrutia, Jose M., and Carmen Echebarria. 2007a. Regional network for quality promotion: A case-study of the Basque Country. *European Planning Studies* 15 (3): 429–451.

———. 2007b. A new internet driven internationalisation framework. *The Service Industries Journal* 27 (7): 923–946.

———. 2010. Social capital, research and development, and innovation: An empirical analysis of Spanish and Italian regions. *European Urban and Regional Studies* 17 (4): 371–385.

BBVA Foundation. 2007. El capital social en España y los países de la OCDE. *Cuaderno de Divulgación 3*. www.grupobbva.com0.

Beugelsdijk, Sjoerd, and Sjak Smulders. 2003. Bonding and bridging social capital: Which type is good for economic growth? In *The cultural diversity of European unity: Findings, explanations and reflections from the European Values Study,* edited by Wil Arts, Jacques Hagenaars, and Loek Halman. Leiden-Boston: Brill.

Beugelsdijk, Sjoerd, and Ton van Schaik. 2005a. Differences in social capital between 54 Western European regions. *Regional Studies* 39 (8): 1,053–1,064.

———. 2005b. Social capital and growth in European regions: An empirical test. *European Journal of Political Economy* 21 (2): 301–324.

Borgatti, Stephen P., and Pacey C. Foster. 2003. The network paradigm in organisational research: A review and typology. *Journal of Management* 29 (6): 991–1,013.

Chen, Xiao, Philip B. Ender, Michael Mitchell, and Christine Wells. 2003. *Regression with Stata.* Los Angeles: UCLA Academic Technology Services, Statistical Consulting Group. www.ats.ucla.edu/stat/stata/webbooks/reg/.

Coleman, James S. 1990. *Foundations of social theory.* Cambridge, MA: The Belknap Press of Harvard University.

Collinson, Simon. 2000. Knowledge networks for innovation in small Scottish software firms. *Entrepreneurship and Regional Development* 12 (3): 217–244.

Cooke, Philip, Nick Clifton, and Mercedes Oleaga. 2005. Social capital, firm embeddedness and regional development. *Regional Studies* 39 (8): 1,065–1,077.

Dakhli, Mourad, and Dirk De Clercq. 2004. Human capital, social capital, and innovation: A multi-country study. *Entrepreneurship and Regional Development* 16 (2): 107–128.

Echebarria, Carmen, Jose M. Barrutia, and Itziar Aguado. 2009. The ISC framework: Modelling drivers for the degree of Local Agenda 21 implantation in Western Europe. *Environment and Planning A* 41 (4): 980–995.

Fleming, Lee, and Matt Marx. 2006. Managing creativity in small worlds. *California Management Review* 48 (4): 6–27.

Florida, Richard. 2003. Cities and the creative class. *City & Community* 2 (1): 3–19.

Florida, Richard, Robert Cushing, and Gary Gates. 2002. When social capital stifles innovation. *Harvard Business Review* 80 (8): 20.

Fountain, Jane E., and Robert D. Atkinson. 1998. *Innovation, social capital, and the new economy: New federal policies to support collaborative research.* Washington, D.C.: Progressive Policy Institute.

Glaeser, Edward L., David I. Laibson, and Bruce Sacerdote. 2002. The economic approach to social capital. *Economic Journal* 112 (8): 437–458.

Griliches, Zvi. 1995. R&D and productivity: Econometric results and measurement issues. In *Handbook of the economics of innovation and technological change*, edited by Paul Stoneman. Oxford: Blackwell.

Hauser, Christoph, Gottfried Tappeiner, and Janette Walde. 2007. The learning region: The impact of social capital and weak ties on innovation. *Regional Studies* 41 (1): 75–88.

Knack, Stephen, and Philip Keefer. 1997. Does social capital have an economic pay-off? A cross-country investigation. *Quarterly Journal of Economics* 112 (4): 1,251–1,288.

Landry, Réjean, Nabil Amara, and Moktar Lamari. 2002. Does social capital determine innovation? To what extent? *Technological Forecasting and Social Change* 69 (7): 681–701.

Leenders, Roger Th. A. J., Jo M. L. van Engelen, and Jan Kratzer. 2003. Virtuality, communication, and new product team creativity: A social network perspective. *Journal of Engineering and Technology Management* 20 (1–2): 69–92.

Maskell, Peter. 2001. Knowledge creation and diffusion in geographic clusters. *International Journal of Innovation Management* 5 (2): 213–238.

Mcfadyen, M. Ann, and Albert A. Cannella Jr. 2004. Social capital and knowledge creation: Diminishing returns of the number and strength of exchange relationships. *Academy of Management Journal* 47 (5): 735–746.

Nahapiet, Janine, and Sumantra Ghoshal. 1998. Social capital, intellectual capital, and the organisational advantage. *Academy of Management Review* 23 (2): 242–266.

OECD (Organisation for Economic Co-operation and Development). 2001. *Measuring capital: A manual on the measurement of capital stocks, consumption of fixed capital and capital services*. Paris: Organisation for Economic Co-operation and Development.

——. 2002. *Frascati manual: Proposed standard practice for surveys on research and experimental development; the measurement of scientific and technological activities*. Paris: Organisation for Economic Co-operation and Development.

Pérez, Francisco, Lorenzo Serrano, Vincente Montesinos, and Juan Fernández de Guevara. 2006. *Measurement of social capital and growth: An economic methodology*. Working Paper 4, BBVA Foundation.

Powell, Walter W., Kenneth W. Koput, and Laurel Smith-Doerr. 1996. Interorganizational collaboration and the locus of innovation: Networks of learning in biotechnology. *Administrative Science Quarterly* 41: 116–145.

Putnam, Robert D. 1995. Bowling alone: America's declining social capital. *Journal of Democracy* 6 (1): 65–78.

Schneider, Gerald, Thomas Plümper, and Steffen Baumann. 2000. Bringing Putnam to the European regions: On the relevance of social capital for economic growth. *European Urban and Regional Studies* 7 (4): 307–317.

Shan, Weijan, Gordon Walker, and Bruce Kogut. 1994. Interfirm cooperation and startup innovation in the biotechnology industry. *Strategic Management Journal* 15 (5): 387–394.

Storper, Michael. 1995. The resurgence of regional economies, ten years later: The region as a nexus of untraded interdependencies. *European Urban and Regional Studies* 2 (3):191–221.

Trajtenberg, Manuel. 1990. *Patents as indicators of innovation: Economic analysis of product innovation.* Cambridge, MA: Harvard University Press.

Uzzi, Brian, and Jarrett Spiro. 2005. Collaboration and creativity: The small world problem. *American Journal of Sociology* 111 (2): 447–504.

Zak, Paul J., and Stephen Knack. 2001. Trust and growth. *Economic Journal* 111 (470): 295–321.

Zheng, Wei. 2010. A social capital perspective of innovation from individuals to nations: Where is empirical literature directing us? *International Journal of Management Reviews* 12 (2): 151–183.

2

Social Innovation, Locational Capital, and Informality in Latin American Cities

PEDRO ABRAMO and ARANTXA RODRÍGUEZ

This chapter discusses some dimensions of the emergence, consolidation, and transformation of what is considered one of the most important social innovations in Latin American cities: the rise of the informal city. In Latin America, the practice of collective land occupations—spontaneous or organized—and their transformation in a popular *habitat* can be considered an important social innovation of the poor in accessing urban land. The "founding act" of land occupation must allow the emergence of other enabling social innovations for the production of basic goods, such as urban infrastructure and collective services, required for the consolidation of that space and its transformation into a popular territory. Self-production of housing and urbanization processes, initiated by a communitarian-type collective action transforms occupied land without urban infrastructure into an informal popular settlement that is effectively the territory (habitat) of social reproduction of the poor. That territory is the result of social innovations that accumulate and gain strength through time. These innovations are diffused and transformed into recurrent practices of the urban poor in the production of their daily urban life spaces. Now, contrary to what early institutionalism authors (Commons 1924) expected, those practices do not transform the legal system that remains illegal or informal and, consequently, "local authorities" are set up to mediate in conflicts and practices in informal settlements. Informal cities are, therefore, regulated by their own institutions that often operate on the margins of the state. At the same time, the reproduction of social and economic life in those settlements depends on reciprocity and redistribution economies of a Polanyian type, in which the market is viewed as an auxiliary economy (Polanyi 1944).

Building upon this characterization of the institutional dynamics of informal settlements, this paper discusses the emergence of a land market that is soon transformed into the primary coordinating mechanism of residential mobility, replacing both occupation processes and/or reciprocity economies as the primary regulatory means of residents' access and exit from these territories. However, the market that evolves within these informal settlements does not abide by the formal institutions that regulate market transactions. Every market transaction involves an exchange of property rights (Commons 1924), but other rights such as commercial, urbanistic or environmental rights also rule market transactions. In the case of the informal land market, land or housing transactions are carried out without any regard for property, urbanistic, or even environmental rights. Thus, land markets in irregular settlements are of an informal nature; that is, the regulatory institutions of market transactions are of a different nature from those integrated in the legal system of the state, the formal market. But because all markets require institutions to regulate contracts, conflicts, or in the case of housing the continuity of the informal sequence chain as coordinating mechanism, the informal market integrates those institutions that were built up socially during the process of territorial consolidation. In other words, the informal institutions established on the basis of reciprocity and distributive economics are now "sucked up" by the informal land market that transforms them in its regulatory institutions.

This chapter examines two dimensions of the conversion of social innovations of the informal city into market valorization elements. The first dimension relates to the individualization of innovations in the form of location preferences and is developed in the following section. Here, we discuss some aspects of the construction of familial rationalities in the context of transforming a social innovation in its market and monetary expression. A second dimension of this process of "sucking up" of innovations is analyzed from the viewpoint of the "locational capital" of the poor. In this case, a social innovation—in this case, the informal habitat—is converted into capital as a result of changes taking place in the intraurban structure that valorize new accessibility factors. This dimension is taken up below.

Location Preferences of Slum Dwellers

There is substantive empirical evidence on the increasing significance of the informal real estate market in determining access to land in slums. There is also considerable evidence that housing prices in the informal real estate market tend to be manifestly higher than prices in the formal market. This trend poses a challenge to prevailing views that competition dynamics between formal and informal markets would result in lower prices in the informal market because of more attractive alternatives in the formal one at equivalent prices. Indeed, recent research reveals that, in reality, competition between the formal and informal real estate markets does not take place.[1] In this section, we attempt to provide some explanation for this apparent paradox through an analysis of the informal real estate market and investment strategies of slum dwellers. In order to explain high(er) housing prices in slums, we examine the links between the labor market and the real estate market and argue that the informal character of the first acts as a barrier to entry in the second's formal market.[2] Given the informality of the labor market and the relative rigidity of housing supply in the real estate market in slums, barriers to entry in the formal real estate market lead to the formation of a "rationed market"[3] that tends to respond to demand fluctuations with marked price variations.

Over the last thirty years, the characteristics of both labor markets and real estate markets have radically transformed the strategies of poor families to gain access to urban land and housing. Research on employment informality and real estate informality has revealed the extraordinary social heterogeneity among families drawing their incomes from means other than regular wages and/or living irregularly or illegally

---

1. Preliminary results of field research about the real estate market (Abramo 1998, 1999) are strongly conclusive as to the preferential choice of real estate property in the slums.

2. The locational terms of this relationship are poorly explored because most residential location models follow the William Alonso (1964) and Masahisa Fujita (1989) hypothesis of central location of the labor market. Likewise, the majority of labor market models are nonspatial (Moyi 2006, 1). More recently, Yves Zenou (1996) has tried to promote an articulation between labor market segmentation theories and the residential location models to attempt an explanation of the dynamics of urban social-spatial structuring.

3. The new Keynesian concept of rationed market is used here. See also Mankiw and Romer (1992); and Dixon and Rankin (1995).

according to urban legislation. This social heterogeneity of the poor was revealed recently by a study on slums based on census-tract sections (Preteceille and Valladares 2000). Likewise, preliminary results from research on real estate markets in Rio de Janeiro's slums uncover a pattern of relative internal diversity among the city's slums as well as substantial internal diversity within slums, a configuration that provides some basis for explaining processes of real estate market segmentation intraslums.

On the other hand, fieldwork interviews among slum dwellers have also confirmed the significance of two critical factors in shaping residential location preferences of the urban poor: proximity to eventual income sources and neighborhood factors. But interviews of second-generation residents provide interesting variations on residential choice. For the younger generations, the notion of proximity to income sources loses its territorial dimension *strito sensu* in favor of a relationship network dimension; an eventual income prospect would be tied to the variety of opportunities provided by the network (religious, parental, or personal relationships). In territorial terms, this network manifests itself in a more dispersed manner. Thus, an employment opportunity is not necessarily linked to physical proximity to a labor demand source such as a factory, industrial or construction site, or household labor market. A relationship network built around Evangelical churches, for instance, opens up new sources of eventual income completely diffused in the city's territoriality. This feature of greater dependency of eventual work opportunities on social relationship networks is one of the new factors introduced in the decision-making process involved in the location preferences of slum dwellers. Nonetheless, this new social "organic" element must not be overrated; data on income source location reveals that physical proximity factors are still quite relevant, and a significant proportion of family income is obtained in the proximity area.

While territorial proximity is often essential in providing eventual income opportunities, the criterion for assessing the significance of proximity between place of residence and a potential employment source is of a different nature than the usual transport cost reduction factor.[4] A

4. The transport cost reduction factor is defined by neoclassical models as the key explaining variable to residential location choices by the poor. A two-fold transportation dynamic could be currently suggested to be verified. Its first aspect pertains to the movement undertaken by the very poor toward a "superperiphery," very distant from the center. The second aspect deals with the access by the poor to more central areas such as the slums. Being central is for these areas a

good example of this "new" type of locational dependency is observed in the need for drug-traffic activities to recruit their labor force from within the community where their base is located (Zaluar 1994, 1998). Trust and knowledge of the territory are, in this case, determining factors for establishing traffic networks in the slums. Both these traits emerge out of proximity relations and can be secured by recruiting labor from the locality where drug dealers operate. In this case, we are confronted with an eventual source of income for the slum dwellers that requires a proximity relation between place of residence and the development of such activity. Nevertheless, over the last thirty years, the relative importance of proximity relationship has changed significantly.[5]

Recent research on the location of income sources of slum dwellers in Rio de Janeiro (Perlman, Vainer, and Abramo 1999) and São Paulo (Baltrusis 2000) has shown that a significant proportion of residents worked inside the slum itself. Thus, the slum can also be seen as a site in which the concentration of informal services and retail activities generates a resource flux that constitutes an "economic circuit" internal to the slum that feeds the local real estate market. This endogenous economic dynamic is established on the basis of such factors as trust and reciprocity. These factors mediate the interaction between social subjects and are critical components of the dynamics of local markets, explaining its endurance vis-à-vis the formal markets at lower prices. But this endogenous economic circuit acts to reinforce the significance of proximity factors in the choice of residential location. Our research results are still preliminary, but they suggest that empirical studies on the internal economy of slums may open up new horizons in the assessment of urban poverty trajectories in Brazilian metropolises, especially in terms of changes in the locational preferences of the poor and the relation to the location of their (eventual or effective) income sources. Differentiating among these three types of proximity broadens up the horizons for understanding economic

---

determining factor on account of the weight of transportation costs to familial overall income. An informal real estate market thus reveals this process of slum densification.

5. The concept of proximity in slums may be effectively qualified using three different dimensions developed by "proximity economics." Physical proximity relates to a physical propinquity that encourages economic and real estate activity. Another type of proximity is defined not by territorial contiguity but by a political and legal definition that establishes a "new" urban territoriality: the slum. A third type of proximity is "organized proximity," established on the basis of institutionalized relations, informal or tacit networks and hierarchies, or even by sheer force and violence.

phenomena in slums and makes it possible to articulate the dynamics of real estate market and residential mobility in slums with a set of other social practices that make up that specific territoriality.

Transforming Urban Territory and Locational Preferences

Residential location is generally considered one of the most important decisions of poor families. Residential location provides for differential access to sources of employment and income, service and retail centers, collective transportation, and public services associated to the positioning of the slum in the location hierarchy of the city. Over the last thirty years, basic infrastructure developments (water supply, sewer systems, electricity supply) have contributed considerably to improving the living conditions of slum dwellers, a trend ratified both by qualitative and by recently published disaggregated sociodemographic indicators for slum areas (Preteicelle and Valladares 2000).

However, while social indicators express that there has been some progress in absolute terms in slums, the pattern of spatial distribution of public infrastructure and services has moved in the direction of strengthening social differences (Marques 1998). The benefits of public intervention in the local sphere are directly tied to the location of public investments that in the last thirty years have strongly favored high-income areas of the city. Generally, these affluent areas concentrate natural positive externalities of the city that are further advanced by substantial public investments. The resulting effect is an increase in intraurban differences. The fact that some slums are in close proximity to these affluent areas makes it possible for them to absorb part of those externalities, a dynamic that also contributes to interslum differentiation.

Interslum differentiation reflects, to a large extent, the dynamics of change in the locational hierarchy of the formal city. This process may thus be seen as *passive* interslum differentiation, as opposed to *active* differentiation resulting from public sector urban initiatives (often highly selective) and/or self-improvement community actions. These public and/or community initiatives have an effect on the slum's physical structure resulting in localized improvements that lead to differentiation of the built environment of a city's slums. Distinguishing conceptually between these two patterns of interslum differentiation, passive and active, is essential for understanding different dynamics of production of the built environment and their articulation to formal and informal territoriality.

Previous studies (Abramo 2001) have identified the dynamics of spatial structuring as a process of continuous differentiation of the built environment marked by a movement of spatial diffusion of spatial innovations that results in an uniformization countertendency. When inter-slum differentiation dynamics, both active and passive, are taken into account, the process of intraurban structuring becomes more complex. This complexity calls for an analysis that takes into account patterns of continuity/change and appropriation/rejection relations between formal and informal urban areas. These relations determine dynamics of valorization/devaluation of the built environment and its externalities. In this way, the slums will be subject to devaluation as a result of changes in the formal built environment but the latter would also be subject to requalification derived from changes in the built environment of the slum.

It can be said, then, that the built environment of the slum and its externalities are transformed through time on the basis of changes that are taking place in the slum itself. But at the same time, change in the slums occurs also as a reflection of changes in the formal environment in the neighborhood. These transformations occur at different scales (street, district, administrative region, zones, and so on), and their primary vector of change is the formal dynamic of transformation of the built urban environment. However, this vector of change, to the extent that is viewed by the production and consumption agents of the formal structure as a "negative externality" of the city, establishes a "conflictive interaction" with the informal built environment.

Thus, the articulation between formal and informal processes of transformation in space and spatial differentiation (formal and informal) in the city is rather complex, projecting (on the basis of a spatial-temporal analysis) a genuine representation of a kaleidoscopic city.[6] In the first case, transformation dynamics in the slum are a manifestation of a dynamic of production of the urban built environment that results from informal relations (land occupation and self-improvement processes, real estate market activity) and/or public intervention on the part of NGOs. Nonetheless, the characteristics and existing hierarchy among slums also change on the basis of the character of its location within the urban structure. To a large extent, the urban structure of the area surrounding a slum and its position relative to other areas of the city is subject to constant

---

6. Pedro Abramo (1998) presents a discussion on the configuration of the kaleidoscopic city taking into account only the formal processes of production of the built-on environment.

transformation driven by the formal real estate market, public intervention, and other agents intervening in urban materiality.

The relative location of slums in the externalities map of a city, in terms of a hierarchy of attributes between slums (interslum preference relations), is transformed over time. A slum, while maintaining its attributes, may exhibit an improvement/worsening of its relative position due to territorial changes within the city. This reflective quality of slums transforms the slum dwellers' residence into a "locational capital" that is valorized/devalued in the course of time—exactly as in the formal real estate market. Family strategies incorporate this factor in their intertemporal budgetary calculations tailing the evolution of the relative position of their "locational capital" (their real estate property) within the intraurban hierarchy, endorsing, in this way, the potential benefits/losses of eventual territorial shifts of the family residential unit (residential mobility) in the intraurban structure. Therefore, a repositioning of this locational capital may stand for both upward mobility as well as a falling standard of living in a family.

As is known, cities display recurring changes in their urban functionality that bare immediate relation to formal land uses. These alterations in the functionality of urban areas affect the locational preferences of the poor. A good example is provided by changes generated as a result of relocation of factory units within the urban space. The preference that the poor have for accessibility to employment sources is well known and is often put forth to explain the emergence of slums (Abreu 1997).

The functional transformation of an area not only has an impact on the relative position held by districts (and slums) in the city's accessibility chart; it also alters its locational preferences. The slum, however, is an urban territoriality that maintains strong primary relationship ties. The sentiment of identity associated with belonging to a "community" is generally seen by the slum dwellers as one of the primary reasons directing a person's choice to stay in the original dwelling places, as is shown by evidence drawn from the interviews. Familial synergy and friendship bonds sprung from many years' daily experiences within the slum (the neighborhood proximity effect) give rise to exchange relationship processes based on the "Gift"/"Counter-Gift" mechanics that allow the maintenance of an expanded "familial solidarity economy."[7] This solidarity economy plays, within the universe of urban poverty, the very important role of raising/

7. The Gift/Counter-Gift economy is presented by Alain Caillé (1994) and Jacques Godbout (1992).

educating children of families whose members mostly participate in the (formal or informal) work market. Neighbors and relatives temporally assume responsibilities that would normally involve either governmental responsibility or monetary compensation, such as nurseries, caring centers for the aged and disabled, and so on. This Gift is returned at a later time, giving rise to vast, complex Gift/Counter-Gift exchange networks that underlie everyday social relations within the community. Location decisions also involve the evaluation of accessibility to such networks of "solidarity economy" manifest, predominantly, in territorialized form. Withdrawal from the slum may then mean leaving such networks.

Residential mobility among the poor may be thus said to consist of very critical decision-making within poor families' intertemporal strategies, especially given the major "social opacity" of the Gift/Counter-Gift networks. Because they are not organized according to explicit norms (tacit relationship), each such network will have a set of traits peculiar to the particular slum in which they are based. The territorialized aspect of these relations may be therefore said to constitute major instances of "informational asymmetry" in the moment that the decision-making process on familial mobility is started. Families that decide to leave a slum (to sell their real estate property) know that, on leaving, their insertion within a specific solidarity network is not to be transferred to the eventual buyer. Thus the family, on being transferred to another dwelling place, must face the uncertainty of encountering a different relationship network, a process that incorporates a "learning cost" for the assimilation of the new tacit agreement.[8] On taking into account the "territorialized solidarity economy" it can be argued that neighborhood externality presents itself as endowed with huge importance for poorer families, and because it constitutes a positive externality, it will probably be highly considered by the decision-making process on residential mobility.[9]

Another interesting feature of the locational preferences of families refers to the possibility of re-creating within the slum some everyday elements of a rural/small-town character that are impossible to reproduce in other metropolitan locations. Due to the peculiar features of their territoriality, the slums make it possible to maintain diverse temporalities

8. This feature may explain research results in some slums where the largest residential mobility was found to be intraslums (Abramo 1999; Baltrusis 2000).

9. The value of solidarity as a positive externality will also be shown in housing prices in slums.

and "lifestyles" that otherwise tend to be driven away from the city. It is interesting to observe how second-generation immigrants tend to absorb urban culture, forsaking their parents' traditions and habits. However, the slum also propitiates new living styles for the young, who maintain the primary-communitarian feature of social relationships. The slum is both the space of daily life for the parents or grandparents that gather to talk by the little bars (local micro-retail), and the place where the nightlife of the funk parties or rap group gatherings takes place, thus making for youth group[10] formation to which the proximity factor propitiated by the slum represents a strong cohesion element. In this way, slum territoriality is turned into a true locational capital, be it in terms of the improvement of the slum's place within the accessibility hierarchy of the city as a whole, be it in terms of neighborhood externalities (accessibility to solidarity economies) or "urban welfare," making for the coexistence of different lifestyles within the slum territory.

The negative/positive externalities, internal to the slum, add to the externalities of its surroundings in the real estate market devaluation process. Fieldwork carried out on the familial universe of slum dwellers suggests that a house constitutes the chief material investment for poor urban families. However, this house is also a kind of property that has the feature of incorporating or giving individual access to urban externalities produced by collective action. During the last thirty years, these activities in the city of Rio de Janeiro and in the peripheral municipalities led to a very unequal pattern in the distribution of urban services and infrastructure. Unequal spatial distribution is expressed in quantitative as well as qualitative terms through a concentration in some infrastructure- and service-based districts, and also in the fact that networks serving poor urban areas are very different from those serving richer areas. Besides, during the last thirty years, these networks evolved in a very aggressive way.[11]

Within the universe of urban poverty, the traditional relation between residential location and location of the workplace (or probable income source) is still one of the key elements around which to build an explanation of the spatial distribution chart of the poor residences.[12] However,

10. See Zaluar (1998).

11. See Silva (2000), Marques (1998), and the classic by Vetter and Massena (1981).

12. Research carried out in several slums of Rio de Janeiro (Abramo 1998) showed that a high proportion (30 percent) of head of households was employed in districts contiguous to their residences.

the spatial segmentation found among service and public infrastructure networks introduces another variable in the locational decision of the families. Thus accessibility to urban externalities, organized according to a very unequal distribution, becomes an important variable in the qualification of popular housing. Accessibility to public welfare institutions, as well as to natural externalities, is very important in facilitating leisure time, which structures the daily universe of those families. Residential location of the poor is, therefore, an important element in their familial strategies, starting with transportation matters but taking into account many other important elements constituting the familial consumer-goods basket.

Another feature of the poorer families' universe of "utilizable" urban externalities relates to proximity to much richer familial groups. This proximity between the slum and high- or medium-class families is normally seen as a somewhat unquiet relationship, as is often seen in the literature of intraurban structure. The effects of neighborhood externality produced by identity and/or social homogeneity (of an ethnic, cultural, economic sort) of the richer families is constrained when a slum appears in the proximity of their dwelling places. But it is also a factor that intervenes in the formation of slum dwellers' human capital.

Conclusions

The production of informal cities and of the social reproduction spaces (habitat) of popular sectors constitutes the primary social innovation in Latin America's large urban centers in the second half of the twentieth century. In this chapter, we have examined some critical aspects related to the emergence, consolidation, and transformation of these innovations, highlighting the risks posed by their appropriation by a market logic, notably in relation to the increasing monetarization of key community traits and interpersonal reciprocity dynamics within land markets. Notably, we have addressed two aspects of this process of market appropriation of social innovations in informal settlements. The first refers to the process of absorption by means of the conversion of the fundamentally collective character of social innovations to an individual value or dimension. The construction of the locational preferences of families is one way in which this individual valorization of externalities generated in the collective process of production of the informal habitat takes place. And it is precisely those locational preferences of the poor that are expressed in the informal land market, sold by those who hold them and transferred through the market.

The second aspect of this appropriation by the market of innovations of production of the popular habitat relates to the relation between that informal territory and the production of the formal city. The location of that socio-spatial innovation and its characteristics in terms of infrastructures and services should not be considered in absolute terms but rather in relational ones. The territorial location of the informal settlement and its valuation within the hierarchy of accessibilities changes with the process of urban transformation and carries within it the virtual capacity to improve (or degrade) its relative position in functions of the vectors and intraurban characteristics of the city's growth. From this perspective, the location of a social innovation over time can be transformed into a locational capital for its residents. A collective process of production of an informal habitat, the social innovation, is transformed into a virtual capital, a "locational capital" that incorporates a monetary dimension in the informal land market. The relational dimension is key to transforming the material output of collective action into an individual value of a relational character because a piece of land, a plot (a piece of the territory of innovation), only has monetary value relative to its position in the urban accessibility hierarchy.

The two dynamics analyzed show the same dimension of market individualization of material outcomes (self-built housing and self-urbanized spaces) and/or social (relations and reciprocity and vicinity economies) or collective actions whose starting point is a social innovation. Here we are confronted with a paradox: a social innovation that is developed in the process of a collective breakup with the market is then appropriated and turned into a commodity to be traded in the informal market. This paradox makes it possible to recognize that the trajectory of a social innovation may at one point be altered and sucked up by a logic of individualization of interests and its mercantilist manifestation as a market exchange. This contingency warns against absolutist interpretations of social innovation and recommends a thoughtful evaluation of normative applications in public policies and strategies.

References

Abramo, Pedro. 1998. *Impacto do Programa Favela-Barrio no mercado inmobiliario de favelas da cidade do Río de Janeiro*. Research report, prelimenary version. Rio de Janeiro: IPPUR-UFRJ.

——. 1999. *A dinâmica do mercado imobiliário e a mobilidade residen-*

cial nas favelas do Rio de Janeiro: Resultados preliminares. Research report. Rio de Janeiro: IPPUR-UFRJ.

———. 2001. *A cidade caleidoscópica.* Rio de Janeiro: Ed. Bertrand.

Abreu, Mauricio. 1997. *Evolução urbana do Rio de Janeiro.* 3rd ed. Rio de Janeiro: IplanRio/Pref.

Alonso, William. 1964. *Location and land use.* Cambridge, MA: Harvard University Press.

Baltrusis, Nelson. 2000. A dinâmica do mercado imobiliário informal na Região Metropolitana de São Paulo: Um estudo de caso nas favelas de Paraisópolis e Nova Conquista. Masters diss., PUC-Campinas.

Caillé, Alain. 1994. *Don, intérêt et désintéressement. Bourdieu, Mauss, Platon et quelques autres.* Paris: Editions La Découverte/Mauss.

Commons, John R. 1924. *Legal foundations of capitalism.* New York: Macmillan.

Dixon, David, and Neil Rankin. 1995. *The new macroeconomics: Imperfect markets and policy effectiveness.* Cambridge: Cambridge University Press.

Fujita, Masahisa. 1989. *Urban economic theory: Land use and city size.* Cambridge: Cambridge University Press.

Fujita, Masahisa, Paul R. Krugman, and Anthony J. Venables. 1999. *The spatial economy.* Cambridge, MA: MIT Press.

Godbout, Jacques. 1992. *L'esprit du don.* Paris: La Decouverte.

Mankiw, Gregory, and David Romer. 1992. *New Keynesian economics: Imperfect competition and sticky prices.* Cambridge, MA: MIT Press.

Marques, Eduardo. 1998. Infra-estrutura urbana e produção do espaço metropolitano no Rio de Janeiro. *Cadernos IPPUR* 12: 2.

Moyi, Tomoya. 2006. "Monocentric versus polycentric models in urban economics." Kier Discussion Paper Series, Kyoto Institute of Economic Research, Discussion Paper no. 611 (January). www.kier.kyoto-u.ac.jp/~mori/papers/DP611.pdf.

Perlman, Janice, Carlos B. Vainer, and Pedro Abramo. 1999. *A dinâmica da pobreza urbana e sua relação com as políticas públicas: Trajetórias de vida em três comunidades do Rio de Janeiro 1969–2000.* Relatório de Pesquisa. Rio de Janeiro: Banco Mundial/Mega Cities/IPPUR-UFRJ.

Polanyi, Karl. 1944. *The Great Transformation.* New York and Toronto: Farrar and Rinehart.

Preteceille, Edmond, and Licia Valladares. 2000. Favela, favelas: Unidade ou diversidade da favela carioca. In *O futuro das metrópoles*, edited by Luiz Cesar de Queiroz Ribeiro. Rio de Janeiro: Revan.

Silva, Roberto. 2000. A conectividade das redes de infra-estrutura e o espaço urbano de São Paulo. In *O futuro das metrópoles*, edited by L. Ribeiro. Rio de Janeiro: Revan.

Vetter, David M., and Rosa M. Massena. 1981. Quem se apropria dos benefícios líquidos dos investimentos do Estado em infra-estrutura? In *Solo urbano: Tópicos sobre o uso da terra*, edited by L. Machado. Rio de Janeiro: Zahar.

Zaluar, Alba. 1994. *Cidadãos não vão ao paraíso*. Rio de Janeiro: Escuta.

———. 1998. Crime, medo e política. In *Um século de favela*, edited by Alba Zaluar and Alvito Marcos. Rio de Janeiro: FGV.

Zenou, Yves. 1996. Marché du travail et économie urbaine: Un essai d'intégration. *Revue Economique* 47 (2): 263–288.

Zenou, Yves, and Tony E. Smith. 1995. Efficiency wages, involuntary unemployment and urban spatial structure. *Regional Science and Urban Economics* 25 (4): 547–573.

3

The Precautionary Principle: Regulation Needed to Confront the Risks Associated with Nanotechnology

Iñaki Lasagabaster Herrarte and Alejandro Arizkun Cela

Translated by Lauren DeAre

The precautionary principle is the concept of adopting certain measures or rejecting others in order to avoid negative consequences. With the precautionary principle, the certainty of the negative consequences may not be scientifically proven or the likelihood of their occurrence is not known. In this sense, precaution goes beyond the prevention principle, which proposes taking actions to avoid negative outcomes.

The European Environment Agency (EEA 2001, 192) has categorized conceptual approximations to these differences according to their level of uncertainty and existing probability:

- *Prevention* is defined as actions taken to avoid risks and their level of known probability. As an example, the agency cites the effects of asbestos dust on well-known respiratory illnesses since 1965.
- *Precautionary prevention* is when identified problems are avoided in favor of those in which the certainty of consequences is unknown. The attempt, therefore, is to avoid potential risks. The effect of antibiotics present in animal feed on humans is cited as an example. The negative impact of the antibiotics is known, but the probability that they will come into contact with human organisms is unknown.

- *Precaution* is when the existence of risks with unknown probability is suspected, but their existence has not been scientifically proven. An example is the damaging effects of chlorofluorocarbon compounds prior to 1974, due to their high levels of permanence and accumulation.

Therefore, the precautionary principle introduces an additional preventive measure. It attempts to avoid certain problems that, once they've occurred, could have irresolvable consequences due to their serious or irreversible nature. They are worth avoiding even though precaution could be expensive in terms of resources and renunciations. In this way, the principle presents a different approach to decision-making in uncertain situations. Instead of waiting for scientific proof of suspected effects to take compensatory action, it proposes avoiding those actions suspected to be harmful or proceeding cautiously until effects can be confirmed. With this approach, the conditions, terms and forms in which the precautionary principle is acceptable must be precisely defined.

The path toward innovation in nanotechnology conforms to the "linear model" of innovation, starting with scientific investigation as the method for reaching a goal.

Nanotechnologies have brought about the existence of products with a multifunctional architecture that can be used for very diverse purposes. At the same time, the consequences of using these products are completely new. Substances or products with dimensions on the nanoscale have effects on health and the environment that are not the same as the effects of the substances in their normal state. Nanoparticles approaching atomic-scale dimensions can enter parts of the body that the substances in their natural states cannot access. The dangers are obvious, as titanium dioxide and zinc nanoparticles have recently shown. In other cases, the use of nanoparticles has been linked to respiratory illnesses.

How to define the application of nanomaterials is a very important legal issue. For any regulation, accurate identification of the regulated material is critically important. In the nano world, this definition is problematic in and of itself. To be considered a nanomaterial, it must have an "approximate" size less than one hundred nanometers. This refers to a dimension, not to an exact size.

Along with the various scientific disciplines, the study of nanotechnology must also take the fields of philosophy and law into account. Rights to health and an adequate environment must also be considered. Despite their importance, nanomaterials are not governed by specific regulations but are covered by other, more general regulations. This chapter will address this issue, particularly within the European Union (EU).

The Importance of the Precautionary Principle in a Crowded World

Because we humans have continued to increase our power on the planet through far-reaching expansion, the precautionary principle is of particular importance. Human impact puts the dynamic balances that govern the planet at risk and could cause irreversible consequences for the survival of human civilization.

Jorge Riechmann (2005) has organized the basic elements of the situation created by forceful human expansion as follows:

1. Humans have filled the world. Faced with this problem of scale, we must propose a *demand-based management principle* that would allow us to adapt our actions to the planet's physical limitations.
2. An inadequate technosphere that adheres to linear growth principles, use of nonrenewable energies, growing residue emissions, destruction of biodiversity, and so on. This problem of design conflicts with the *biomimetic principle*, or the technology that mimics nature. As demonstrated by solar energy, this principle completes cycles by reusing materials and strengthening diversity.
3. Development that features noticeable physical inefficiency in energy and material usage. Making economic efficiency the priority in development of our societies has caused a lack of conservation of energy and inexpensive materials. This has led to a problem of efficiency that we must counteract with an *eco-efficiency principle*, or efficiency in physical terms.
4. Implementation of a powerful technoscience system that has attained its own expansive and irreflexive force over the goals of technoscience activities. This problem, which has been called a Faustian problem, must be controlled by the *precautionary principle.*

In the scenario described, risks are multiplied and require special caution, various adjustments and a prudent approach wherein the precautionary principle must have a significant role. Many experts have warned of the risk of the collapse of human civilization if no adjustments are made and if we continue to push the physical limits of the planet.

In this context, Robert Constanza (1989) proposed the following approach to (1) the relevance of precautionary principle application, and (2) consideration of technology policies as measures that can prevent humanity's hypothetical collapse. There is no doubt that certain technological improvements could help improve some aspects of this excess, but in order to avoid it, are technological measures the solution? Will the effects of applied technologies be enough to avoid a collapse? Will technological improvements come fast enough to postpone that collapse? Future developments are regarded with a high level of uncertainty.

Let's imagine an environment in which those who are optimistic about technology are correct. In the future, a fast technological evolution occurs and is properly directed, thereby avoiding a collapse. On the other hand, we also have to consider a possibility foreseen by a large number of experts, those who we'll call pessimists. In this scenario, the direction, scope, or speed of technological advancement is ineffective or insufficient, and cannot prevent a collapse. In this situation, what would the advantages and disadvantages of enacting policies of one kind or another be? Policies that show confidence in human abilities to find new processes, resources, and products, and those that trust in technological improvements to resolve problems are considered technologically optimistic. Pessimistic policies are those that do not have confidence in the certainty that we'll find those improvements, and propose demanding other measures, in particular, self-restraint in our use of the environment. Four situations result from this approach, included in table 3.1.

Table 3.1. Hypothetical effects derived from different politics

	Optimistic scenario	Pessimistic scenario
Optimistic politics	**positive**	**disastrous**
Pessimistic politics	**moderate**	**tolerable**

Within the current context of uncertainty and ignorance, the application of technologically optimistic policies could have disastrous effects. Technologically pessimistic policies could prevent disaster. If incorrect, however, certain improvements could be lost. The precautionary principle proposes the adoption of an approach that would minimize irreversible losses. The price to pay would be the improvements lost if technology behaves as described in the optimistic scenario. It would involve implementing technological advances as they were created, but not trusting *a priori* that they alone will solve all problems.

The Conditions and Fields of Application of the Precautionary Principle

When setting aside scientifically arrogant positions that attribute an absolute or quasi-absolute value to scientific knowledge and instead accepting the limited, changing and temporary nature of said knowledge (Morin 1984), a question arises concerning the conditions for the application of this principle. With a strict application of the precautionary principle, meaning if we demand a definitive scientific demonstration of an absence of risk, a paralysis of human activity could occur. The precautionary principle is an approach to apply in uncertain situations.

Roberto Andorno (2004) has categorized the basic elements of the debate over the conditions for application of the precautionary principle, and this debate is far from over.

One of the first points that Andorno makes (2004, 8) is that the precautionary principle should not be considered "a resistance to progress or a halt to technological innovation." It does not try to fight technological advancement. It doesn't attempt to reverse our social dynamics or revert our societies back to allegedly primitive and happy ones based on the values of the "noble savage."However, a reorientation of the path that technological advances are taking is necessary. Advances must be subordinate to human development objectives, as they were previously designed.

Another caution that Andorno (2004, 72) sets forth is that "this principle does not necessarily imply refraining from certain technologies . . . in many cases, the precaution would be limited to intermediary measures, such as: promoting further depth in scientific studies in order to have a more complete idea of the magnitude of risk; requiring use of consumer information labels; closely monitoring products and their effects; encouraging the search for safer alternatives for the population."

The debate surrounding the conditions of application for the principle must at least include the following questions:

- Situation of uncertainty pertaining to risk
- Scientific evaluation of risk

- Potential for serious and irreversible damage
- Proportionality in actions
- Transparency in actions and participation in decisions
- Reversal of the burden of proof

The health effects of different human actions have been the first wake-up call for the need for a more prudent approach. In 1962, Rachel Carson's pioneering *Silent Spring* warned of the detrimental effects of using artificial chemical substances, including herbicides and insecticides on human and other living beings. Since then, our knowledge of the harmful effects of different xenobiotic substances and artificial radiation used in production processes has increased dramatically. Despite these notable advances in our knowledge of harmful effects, introduction of advances greatly outpaces how fast studies of their effects can be performed. Therefore, in most occasions when measures are implemented to prevent negative outcomes, their irreversible nature makes them impossible to eliminate. Given this information, it seems appropriate that the field of human health should become the first place to implement changes in practices that have traditionally given precedence to productive activity over precaution, because of its importance and due to the seriousness that negative effects can have on human health.

The environment is the second field where these concerns apply. In some cases, the value we give to the living beings and inanimate objects that surround us is derived from internal values attributed to the beings themselves. In other cases, their importance is based on what they can offer to the lives of human beings through interdependence and ecosystems.

Therefore, a field of application for the precautionary principle is one that pertains to the effects that certain products or specific activities have on human health or the environment.

Additionally, there is another field that addresses the grand course of planetary evolution as influenced by human activities. If a risk of human civilization collapse as a result of anthropogenic actions exists, no matter how indefinable the specific time frames and exact characteristics may be, a very large field opens up for reflection on the precautionary principle's role in examining human behavior and measures that could reduce those risks.

However, nanotechnology is the focus of this chapter, and so we must further analyze the role of technology in our societies.

Technoscience and the Precautionary Principle

There's no question that available technological advances have facilitated growth in human ability to confront various problems. On the other hand, these advances have introduced risks and a potential for self-destruction that cast some shadows over technoscientific advancement. Martin Rees (2004, 35) warns of the potential risks that biotechnology and nanotechnology are creating in the twenty-first century. Currently, many natural disasters are related to climate change caused by human activities.

The author classifies several characteristics of technoscientific progress that are of interest here:

- The increasingly autonomous nature of technoscientific advancement with less social control due to parceled knowledge and specialization
- The belief that technology has unlimited capabilities
- Lack of reflection on the consequences that result from the use of products, procedures, or research activities
- The existing belief that science alone can produce beneficial outcomes

In addition, technology is presented as a successful way to solve the problems that technology itself has created. Technoscience[1] has its limits. According to Nicholas Rescher, "We know and should acknowledge the limitations of our cognitive abilities. We cannot, in any justified way, equate reality with what can, in principle, be known by us; nor can we equate reality with what, in principle, we may be able to express with our language. Any physically viable cognitive being can only know a part or an aspect of reality."[2]

The limits of technology are not only tied to our cognitive limitations but also to the entropic nature of the universe. The second law of thermodynamics shows that in time, there's a continual increase of entropy. In an open system, like planet Earth, at the same time that we decrease entropy through use of solar energy, we exchange it for an equal increase of entropy in another part of the universe.

---

1. We use the term *technoscience* to include the growing level of interrelation between science and technology. Advances in science have provided access to increasingly powerful technologies, made more refined tools available, and allowed advancement of scientific knowledge.

2. Cited by Riechmann (2004, 120).

Nanotechnologies have sometimes been presented as technological mechanisms that are able to decrease entropy: They attempt to create order by removing toxic residues atom by atom that are created by industrial expansion and are used to create useful and durable products. From these projects, nanorobots that cost millions to manufacture will be used to produce this antientropic reconstruction. The applications for these nanotechnologies will be very diverse, for example:

- Medical treatments (nanoparticles that can attack cancer or nano-bombs that allow therapies to be delivered to very precise locations)
- Bionic combinations of living materials and nanotechnologies used to develop more advanced living beings
- Elimination of old age through reconstruction of the body's elemental atoms with new materials
- Ending hunger with nano food products
- Elimination of contamination through atom-by-atom absorption[3]

But once again, thermodynamics puts this alleged illusion of grandeur in its place. The massive development of these technologies demands a great amount of energy, an activity that creates disorder.

Other risks of the use of these nanotechnologies have been described (Reichmann 2004):

- Those derived from the social control that may occur as a result of monopolistic control in the hands of a few
- Health problems resulting from the penetration of nanoparticles into live cells, thereby altering their function

Application of the precautionary principle in the management of nanotechnologies is particularly relevant in the face of these potential risks.

The Precautionary Principle and Its Application in Dealing with the Risks Associated with Nanotechnology

A report issued by the European Parliament describes the weakness of the European Commission's arguments pertaining to the legal regulation of nanomaterials. The report states that the issue falls under the applica-

3. Taken from a summary of various publications by Riechmann (2004, 129).

tions of already existing regulations. This approach is unsustainable. The deficiency of current regulation is indisputable. The need for regulation of nanomaterials to protect health and the environment is unavoidable. Nanotechnologies bring to light the consequences, in all their brutality, that these risks present to society. Our society is facing the risks of an issue that public authorities need to confront, without letting market forces pressure its abandonment. The lack of regulation and answers to the challenges presented by nanotechnologies represents a neglect of responsibility on the part of public authorities. This is where the precautionary principle should be fully employed. Neglect of this principle has exposed the weakness of the system in some of its most basic elements.

The precautionary principle has already been defined in an eponymous document created by the European Commission. That document states that the precautionary principle should be considered in the framework of a structured risk analysis strategy that includes three elements, namely: valuation, management, and communication of risk (Commission of the European Communities 2000).

As already described, the precautionary principle must be applied to situations in which uncertainty is not scientifically proven. However, science doesn't have all the answers, which leaves us to face the end of certainty. Nanotechnologies paradigmatically represent the uncertainties produced by technological developments as a result of scientific advancement. A political answer in the form of legal regulation must be provided for questions that arise from the implementation of nanotechnology. Several factors for the application of the precautionary principle in this field should be considered, namely: (1) the way that it could measure risks and make them acceptable, (2) the participation of the population in decisions related to those risks, and (3) the methods or tools that the administration will implement in order to comply with regulations related to the subject.

Experiments and research conducted up until now have provided insufficient data for forming an opinion about the consequences of implementing nanotechnologies. But this does not mean that in the fields of science the consequences of nanomaterials on health and the environment are unknown. Risk that hasn't been subjected to a systematic study from a scientific point of view cannot be accepted. Research data must be comparable and rigorously executed for this systematic study.

Confidence or faith in scientific progress must be limited by the knowledge of its consequences. The risk to society should be a known risk, in the sense that the uncertainty should be situated within set parameters and recognized by the population.

The precautionary principle requires preventative intervention on this topic. In the same way, this principle is linked to sustainability and requires improvements to regulations like those that regulate chemical products subject to approval. According to those regulations, and despite the dangers, a chemical product can be approved if social economic reasons justify its use. The sustainability principle does not allow reasoning of this type, in which economic policies can take precedence over environmental or health protection policies. The same can be said in the case of community participation, where the basic principle must not be hidden behind a screen of experts. The value of expert reports is diminished by the uncertainty that accompanies the political decision of allowing these products to be used. This, along with a basic democratic principle with emphasis on environmental issues, like the one stated at the Aarhus Convention, forces public policies on the subject to focus on public participation. That participation must be accompanied by information and the right to access public documents. Participation in this issue is linked to a fundamental principle: the acceptability of social risk. This acceptability is difficult to produce if it's considered an issue that belongs to researchers or specialized experts only. Debates must extend to include public opinions and especially pubic opinions from sectors that are directly affected by or interested in environmental or health risks (Marini 2007, 13–14). Products used in agriculture should not be approved without the express participation of farmers.

Our last point deals with the legal methods used to implement regulations. The precautionary principle requires an application of strong intervention instruments by public authorities. Those instruments cannot be diluted into ethereal categories that do not clearly and precisely identify which public and private parties are obligated to intervene or what decisions must be made. The resulting instruments must be coercive to implementation. Policy management will require the establishment of a system of approving nanomaterials that go into the market. The legal regulations applicable to nanomaterials should avoid self-regulating procedures or weak decision-making systems where the responsibilities are not clearly defined.

It can be stated that the current European policy on the subject is mistaken. First, the Commission states that the existing regulation is sufficient to guarantee adequate management of nanomaterials. However, in the following line, it states that new regulation is necessary in certain areas such as cosmetics, new foods, and food additives (Commission of the European Communities 2009, 1,468). The European Parliament and the European Economic and Social Commission (EESC) both state that the current existing regulations are not sufficient to guarantee health and environmental safety in dealing with nanomaterials. If these issues are not properly attended to and risk studies are not performed with greater investment, there will be an unavoidable price to pay in terms of future civil responsibility and environmental cleanup.

References

Andorno, Roberto. 2004. Validez del principio de precaución como instrumento jurídico para la prevención y gestión de riesgos. In *Principio de precaución, Biotecnología y Derecho*, edited by Romeo Casabona and Carlos María. Bilbao-Granada: Fundación BBVA-Diputación Foral de Vizcaya-Comares.

Carson, Rachel. 1962. *Silent spring*. New York: Houghton Mifflin.

Commission of the European Communities. 2000. Communication from the Commission on the Precautionary Principle. Brussels, February 2. COM (2000) 1. http://eur-lex.europa.eu/LexUriServ/site/en/com/2000/com2000_0001en01.pdf.

——. 2009. Nanosciences and Nanotechnologies: An Action Plan for Europe 2005–2009. Second Implementation Report 2007–2009. Communication from the Commission to the Council, the European Parliament and The European Economic and Social Committee. SEC (2009) 1468: COM (2009) 607 final. http://eur-lex.europa.eu/LexUriServ/LexUriServ.do?uri=COM:2009:0607:FIN:EN:PDF.

Constanza, Robert. 1989. What is ecological economics? *Ecological Economics* 1: 1–7.

EEA (European Environmental Agency). 2001. Late lessons from early warnings: The precautionary principle, 1896–2000. Environmental Issue Report 22. Luxembourg: Publications Office of the European Union. www.eea.europa.eu/publications/environmental_issue_report_2001_22.

Marini, Luca. 2007. OGM, precauzione e coesítenza: Verso un apporccio (bio)políticamente corretto? *Rivista Giuridica dell'ambiente* 1: 1–15.

Morin, Edgar. 1984. *Ciencia con consciencia.* Barcelona: Anthropos.

Rees, Martin. 2004. *Nuestra hora final: ¿Será el siglo XXI el último de la humanidad?* Barcelona: Crítica.

Riechmann, Jorge. 2004. *Gente que no quiere viajar a Marte: Ensayos sobre ecología, ética y autolimitación.* Madrid: Libros de la Catarata.

——. 2005 ¿Cómo cambiar hacia sociedades sostenibles? Reflexiones sobre biomímesis y autolimitación. *Isegoría* 32: 95–118.

4

New Technologies in the Globalized Economy: Patents as Tools for Measuring and Comparing Innovation Systems

Itziar Urizar and Virginia Fernández

Translated by Jennifer R. Ottman

Technological innovation took on special significance as a consequence of the crisis of the 1970s and the competitive weakness shown by the developed economies. New technologies, together with market globalization and the development of the *knowledge economy*, have now put us in the position of facing a *new economic geography*.

Analyzing patents as tools for measuring and comparing innovation systems around the globe will enable us, in the course of this chapter, to reveal the configuration of the new world technological map. Prior to the detailed study of patents as indicators of the globalization of knowledge, it will be useful to address the conceptual framework within which this article contextualizes the study of the experiences of India and China, new protagonists of innovation in Asia.

In this regard, a generalized tendency can be detected around the world, starting in the 1980s, to progressively shift the conceptual framework for analyzing experiences of technological development from "national innovation systems" to "regional innovation systems."

Conceptual Framework

The concept of a national innovation system first appeared in the 1980s to describe Japanese activity in the postwar period (Freeman 1987). Its use became progressively generalized in the 1990s as a frame of reference for studying international experiences of technological development (Lundvall 1992; Nelson 1993; Edquist 1997).

Subsequently, a large part of the literature on national innovation systems moved on to more in-depth exploration of the regional and local levels and to studying the potential of territorial innovation systems for promoting the use of *social capital* and institutional networks in order to further their capacity for adaptation in an increasingly globalized economy (Cooke, Heidenreich, and Braczyk 2004, 1–18).

For Lars Coenen and Bjorn Asheim (2006), *the regional innovation system is considered to be the widest appropriate intellectual framework for guiding public action.* This normative character, together with the systemic and evolutionary vision of these theories of innovation, contributes to providing public authorities with a more concrete framework within which to focus their efforts to support innovation.

Until the 1990s, the study of experiences of technological development made it possible to highlight the significance of R&D investment as a strategic factor. This impression was reinforced by the system of innovation indicators and measurements adopted by the group of OECD countries during the 1950s and 1960s and that would be standardized in the *Frascati Manual* (OECD 1963).

Nevertheless, as the processes of technological development became more complex, it became necessary to consider other qualitative factors, together with these quantitative indicators, taking into account the fact that innovation was acquiring a more systematic and interactive character (von Hippel 1988; Lundvall 1988; Nelson 1993).

Innovation became a social process of producing and diffusing new knowledge, taking place in a variety of institutions as a result of the interaction produced within the framework of a regional innovation system (Lundvall 1992). From this perspective, knowledge and processes of learning are tied to a specific territory (Navarro 2009, 26; Lorentzen 2009, 171–83).

Nevertheless, despite this tendency toward increasing regionalization in studies of innovation systems, the continuing lack of regionalized innovation indicators, as well as market globalization and the intensification of the state's role in creating and maintaining competitive advantage, has meant that national innovation systems continue to be taken as the basic point of reference for economic analysis, especially for conducting comparative analysis on a global scale, as this chapter aims in part to do. In this regard, the emergence of new Asian protagonists on the new global map of innovation, such as China and India, appears to reinforce the adoption of an approach based on national innovation systems as the fundamental analytical framework.

Patents as Indicators of the Globalization of Knowledge

In 1993, the OECD proposed a classification of different productive sectors by level of technology, calculated on the basis of R&D expenditure (OECD 1997). This classification provided a point of departure for analyzing the evolution of productive activity and technological change.

Subsequently, Paul Robertson and Parimal Patel (2007) observed that those countries in which high-technology sectors have the greatest presence are also those that host medium- and low-technology sectors, due to the transversality of the technologies in question and to the diffusion of knowledge throughout the web of production.

During the 1990s, research and the acquisition of knowledge were oriented toward the development of incremental innovations or the production of "synthetic knowledge" determined by observation and facts, associated with the predominance of industries such as machinery, electrical engineering, and automobile manufacturing. Subsequently, the emergence of other cutting-edge sectors such as pharmaceuticals, biotechnology, nanotechnology, and advanced software has corresponded to regimes of "analytical knowledge," independent of facts or experience (Laestadius 1998). In these new sectors, the pursuit of patents has intensified (Asheim and Coenen 2005). That is to say, aside from R&D expenditure, there are other factors, such as patents, licensing agreements, and cooperative relationships, which determine an innovation system's technological capability.

In addition, patents are a significant source of statistical information about the globalization of knowledge. Even if an invention originates in one country, it can be applied in a different territory. That is to say, classifying patents registered in a country by their place of origin is a meaningful indicator of a country's degree of openness to the wider world or of the globalization of its scientific and technological activities.

Figure 4.1 reflects the upward trend in patent applications worldwide between 2000 and 2006 and shows the attractiveness of seeking commercial applications of research as a function of the demand and opportunities offered by each market.

Figure 4.1. Patent applications at the worldwide level (2000–2006)

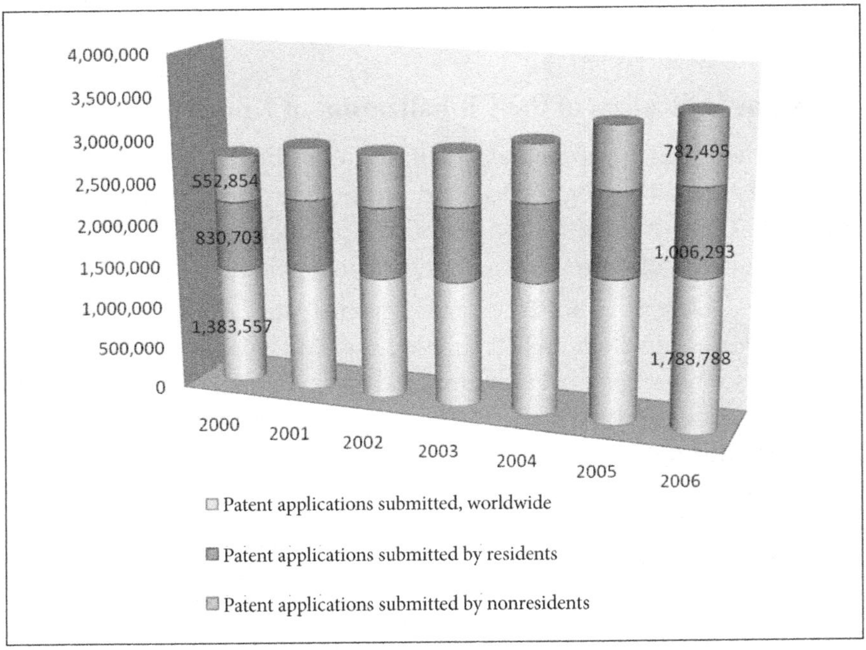

Source: WIPO (2009).

Even if patents do not provide an absolute criterion for analyzing transformations in productive activity, they do contribute, due to their strategic character and to the fact that they are concentrated in cutting-edge activities, to providing some points of reference with regard to the competitiveness of global industry and the changes undergone by innovation systems over the course of time (Gómez Uranga, Etxebarria, and Barrutia 2009, 113).

Table 4.1. Patent applications by technological sectors (2000–2006)

	Japan	USA	Germany	Korea	China	France	Spain	Brazil	India
1. Electricity/electronics									
Electrical appliances, electrical engineering, electrical energy	264,686	92,085	49,477	58,493	24,514	15,091	1,397	1,065	201
Audiovisual technology	268,218	69,780	16,455	65,493	17,818	9,848	550	455	71
Digital communications	197,719	128,177	21,193	79,456	33,527	17,389	908	441	140
Basic communications processes	60,386	85,591	13,989	25,052	24,285	11,748	408	112	116
Information technology	42,388	28,747	7,200	9,085	3,054	3,036	103	48	131
IT management methods	50,958	42,936	3,066	8,339	2,452	1,304	123	153	72
Semiconductors	217,261	78,442	20,223	72,471	12,743	4,881	150	48	26
2. Instruments									
Optics	279,928	63,161	16,918	52,107	130,039	5,993	302	198	65
Measurement	147,329	95,497	44,112	15,430	20,504	11,870	1,053	699	221
Analysis of biological materials	18,210	39,625	8,159	1,664	3,961	3,267	482	138	152
Control	75,964	47,075	17,015	9,189	7,654	6,090	995	655	74
Medical technology	112,685	222,133	49,739	13,893	45,202	20,622	3,927	1,651	2,453
3. Chemistry									
Developed organic products	60,706	115,792	45,389	9,224	15,537	20,850	2,862	546	4,623
Biotechnology	37,252	110,378	19,233	7,652	14,107	8,096	1,548	418	1,089
Pharmaceutical products	43,451	165,330	33,080	7,873	39,318	15,761	3,545	877	4,172
Macromolecular chemistry, polymers	80,038	48,048	25,683	8,092	9,686	6,355	275	269	241
Food chemistry	27,257	26,780	5,813	11,378	17,450	2,872	864	869	646
Materials chemistry	77,928	67,245	32,590	10,402	20,895	6,872	864	869	646
Materials, metallurgy	77,897	30,659	18,119	10,960	19,474	6,964	715	586	464
Surfaces and coatings technologies	94,474	56,161	19,332	9,056	8,312	5,955	575	315	128
Nanotechnology microstructures technology	5,758	4,296	1,492	1,736	885	631	33	12	9
Chemical engineering	73,566	62,937	30,307	10,713	12,881	9,724	1,255	762	448
Environmental technologies	60,261	25,047	13,665	11,344	9,612	5,243	724	388	183
4. Mechanical engineering									
Management	100,002	53,361	27,646	10,517	5,966	9,868	2,088	1,115	138
Tool-making machines	75,586	43,975	30,402	11,556	10,402	6,021	901	522	91
Motors, pumps, turbines	92,695	42,756	39,932	13,476	7,505	9,744	795	850	121
Textile and paper-making machinery	126,016	39,841	24,243	9,661	8,693	4,816	640	449	142
Other specialized machinery	101,012	63,535	32,310	14,873	15,123	11,055	2,129	1,360	228
Thermal processes and apparatuses	55,536	21,019	14,917	19,448	11,460	3,789	706	512	85
Mechanical components	105,285	46,231	49,476	12,056	7,746	11,847	1,129	781	105
Transport	146,516	62,270	69,035	30,512	8,263	22,673	2,525	1,212	140
5. Other sectors									
Furniture, games	95,748	59,112	14,444	19,027	7,838	6,736	1,597	1,189	54
Other consumer products	62,854	43,888	16,527	23,415	10,398	8,618	1,295	1,246	87
Civil engineering	88,120	53,592	27,216	22,964	16,535	11,243	2,843	1,308	76

Source: WIPO (2009).

Table 4.1 lists numbers of patents by field of technology, according to the World Intellectual Property Organization classification, for the period 2000–2006. The countries are given in order of their number of patent applications. Japan comes out ahead of the United States in almost all fields except digital communication, biological materials, medical technology, biotechnology, organic chemistry, and pharmaceuticals. These results are not coincidental because Japanese industrial firms are more competitive than American ones, and the biological and medical sectors have greater entrepreneurial importance in the United States.

Germany surpasses the United States only in transport and mechanical elements, but is surpassed in its turn by Korea in all the electrical engineering and communications fields, in addition to some specific areas in the field of nanotechnology. China surpasses Korea in biotechnology and in the chemical fields.

France, Spain, and Brazil do not present distinctive elements because each of these countries is surpassed by the one before it in the list, and the last is surpassed by India only in the medical and biological area, where India's strong generic pharmaceutical sector must be taken into account.

In terms of intensity, if we compare the ratio of patent applications to GDP, Japan has one of the highest indexes (see figure 4.2).

Figure 4.2. Patent applications (residents) in relation to GDP (in millions of constant US dollars)

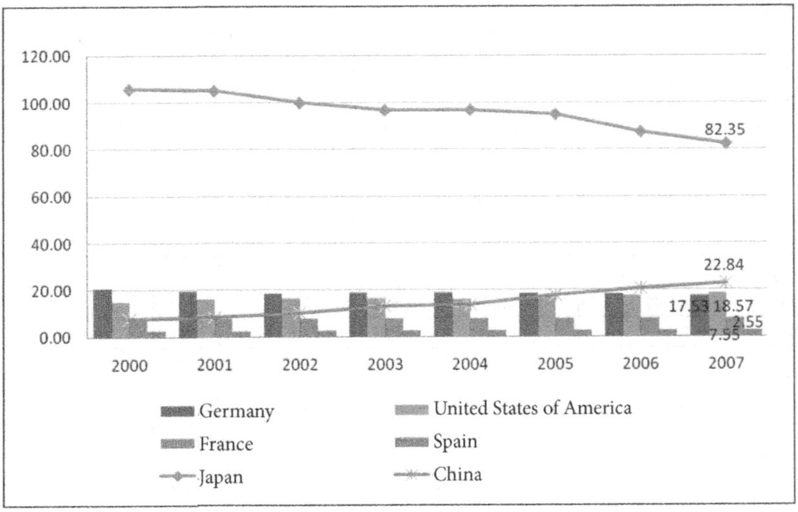

Source: WIPO (2009).

Nevertheless, a notable decrease in the number of Japanese applications is evident, declining from 35.9 percent to 29.1 percent between 2000 and 2006. China, on the other hand, shows a continuous increase since 2000 (see figure 4.3).

Figure 4.3. Patent applications by country of origin

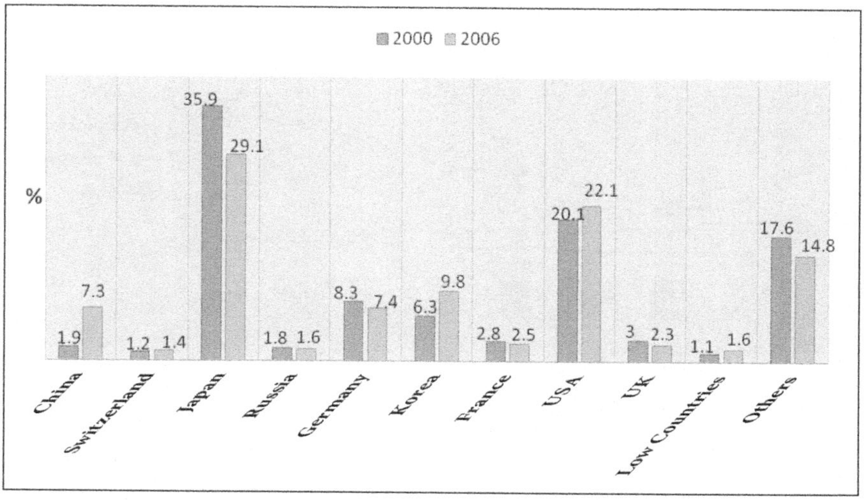

Source: WIPO (2008).

In general, we observe a similar or slightly lower number of applications per resident in 2007 compared with 2000 because the GDP grew more rapidly than the number of patent applications (see figure 4.2).

Worldwide, the number of patent applications per million dollars of R&D expenditure has changed very little since 2000, even decreasing in the case of Germany, and doing so very sharply in the case of Japan. Only China shows a rising index of intensity in relation to R&D expenditure (see figure 4.4).

The current economic and technological map, compared with the one prevalent until the beginning of the 1990s, clearly reveals that more or less striking changes have taken place in many economies around the world.

In the Americas, the United States maintains unquestioned technological leadership. In Europe, Germany continues to appear in a prominent position, although it has not demonstrated the rhythm of innovation that might have been expected from it.

Figure 4.4. Patent applications (residents) in relation to R&D spending (in millions of constant US dollars)

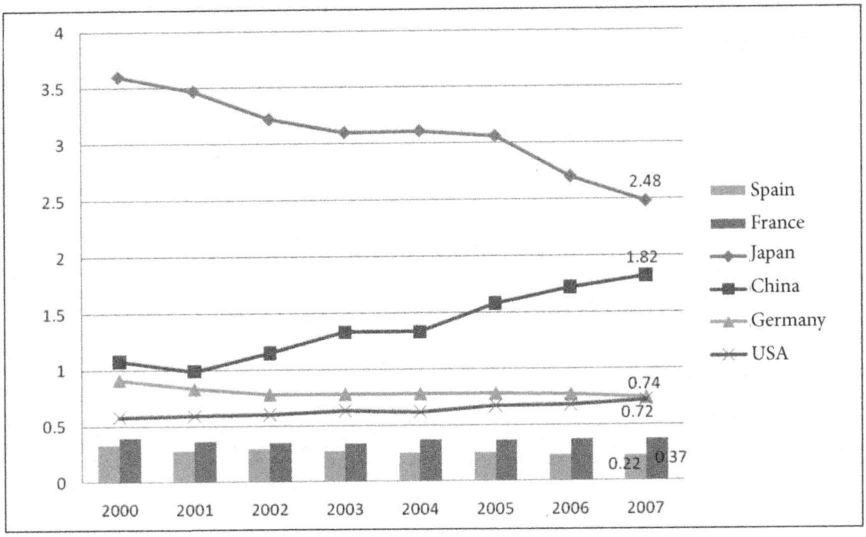

Source: WIPO (2009) and UNESCO.

In Asia, Japan continues to play a significant role. Nevertheless, in contrast with its position of almost exclusive leadership in the previous model of innovation and technological development, economies such as those of China and India, to the analysis of which we dedicate the next section, have attained noteworthy positions in the new model.

Innovation in Asia

The emergence of China and India onto the global economic scene is one of the most significant events of recent years. The enormous growth rates recently accumulated by these countries and the fact that they together account for one-third of the world population make the study of their models of growth uniquely interesting.

If India's growth has revolved around specialization in the service sector, and specifically in information, communications, and software technology, China has chosen to invest fundamentally in industrial development.

Both models are characterized by the state's leading role in promoting innovation. At the same time, the establishment of China's modern innovation system reflects the strong centralization and verticality of an authoritarian economic model based on central planning. For its part, India has found a way to integrate the potential offered by the various levels of its innovation system by including the regional perspective, in which the Indian states enjoy extensive autonomy in developing specific innovation policies.

India: The Case of the Software Industry

Following independence in 1947, the Indian government chose to invest in science and technology as strategic resources for the country's modernization and economic transformation.

The first Science and Technology Plans insisted on the importance of infrastructure development and on training researchers in new technologies. Progressively, the emphasis of these plans shifted to applied research, technology transfer, and knowledge diffusion (Aizawa et al. 2008, 40–41).

At the same time, the fundamental characteristic of technological change in India is that it has taken place chiefly in the sector of information technology and related services, which has received targeted political support to the detriment of other sectors, the innovative capacity of which has not been stimulated to the same extent (Heeks 1996; Parthasarathy 2004).

In 2004, India accounted for about 65 percent of the global market in software and services and about 46 percent of the market in business process outsourcing (Mitra 2007, 61).

This sector has grown very significantly in recent years, as can be seen in figure 4.5.

Figure 4.5. Information and communications technology industry: hardware and software

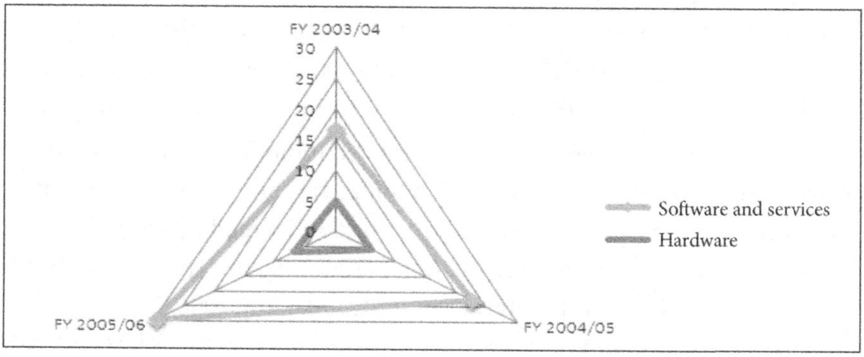

Source: NASSCOM (2006).

In the process of globalization in the software and information and communications technology industries, which began in the 1980s, in addition to an active policy of attracting foreign investment (Desai and Bhalerao 1996) and a trade policy encouraging technology imports and promoting exports (Aizawa et al. 2008, 43–46), two other key elements must be mentioned.

One is an educational policy aimed at providing this sector with specialized scientists and engineers in an attempt to anticipate the demand for human capital. As a result, there are now seven Indian Institutes of Technology of acknowledged international prestige, producing highly qualified engineers and scientists (Sancho and Manchón 2006, 54).

In 1998, 30 percent of Indian Institute of Technology graduates and 80 percent of degree recipients in the field of information sciences went to the United States to continue their studies (Lundvall, Intarakumnerd, and Vang 2006, 70). This "brain drain," the members of which have subsequently gone on to leadership roles at Silicon Valley firms, has had a positive influence since the 1980s on American industry's decisions to outsource information and communications technology to India (Saxenian 2002, 192).

In the second key element, the Department of Electronics has led the way in creating infrastructure oriented toward the development of the software industry (Lundvall, Intarakumnerd, and Vang 2006, 233–34). The Indian government has also contributed to this development in recent decades by creating technology parks through the Software Technology Parks of India (STPI) program. By way of this program and other initiatives aimed at integrating regional capabilities, developing human capital, attracting foreign investment to the territory, and maintaining collaborations and exchanges of information with Indian scientists established abroad, significant sectorial agglomerations have been developed around information and communications technology in areas such as Bangalore.

In effect, one of the keys to India's technological development in information and communications technology has been its success in integrating the capacities of the international system created by globalization, the national system, and its regional innovation systems, to the benefit of the local systems (Lundvall, Intarakumnerd, and Vang 2006, 87). The existence of a federal system of support for regional development has allowed the regional authorities greater room for maneuver in establishing their own policies, with regard in particular to improving the educational system and attracting businesses.

In order to continue taking advantage of its innovation potential and maintain its current rate of growth, India needs to strengthen its R&D capabilities at both national and regional levels, without losing sight of its chief international points of reference. This will mean intensifying its R&D efforts, which in 2006 still accounted for only 0.7 percent of GDP, while China spent almost twice that, 1.4 percent, with the United States spending 2.6 percent and Japan 3.3 percent (OECD 2009).

On the other hand, India has to try to progress toward greater integration of its innovation system in order to improve the commercialization of new knowledge and translate research efforts into commercial applications.

In figure 4.6 we observe moderate growth, denoting India's internationalization efforts in the area of technological development, but we must also still highlight the much higher percentage of nonresident applications from abroad.

Figure 4.6. Patent applications, by patent office, presented by residents/nonresidents (India)

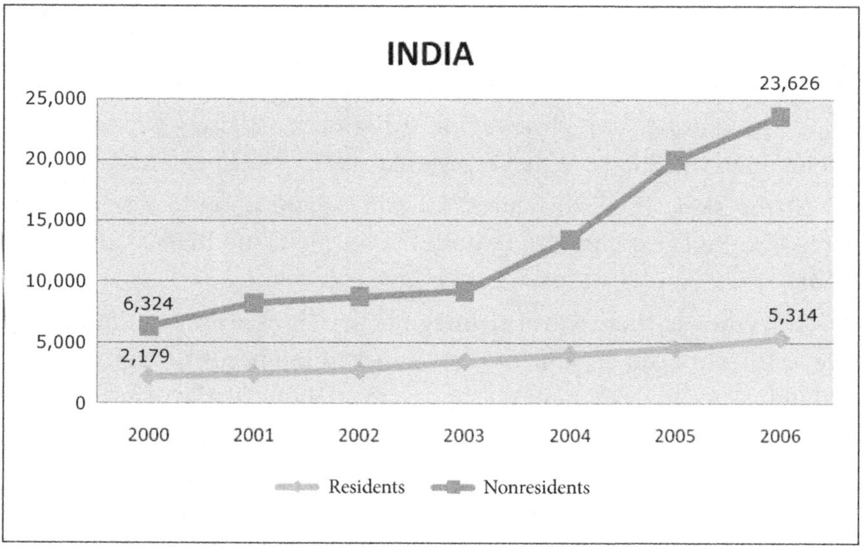

Source: WIPO (2009).

In sum, India has to deploy the resources that will enable the definitive transition of its economy to a modern economy capable of generating new job opportunities for its poorest citizens within a global strategy of the diffusion and utilization of knowledge.

China: The Other Pole of Globalization

The United States and China are currently the two great poles of globalization. China's global economic significance rests on its enormous size and rapid export-driven growth (Aizawa et al. 2008, 76–119), its capital, and foreign investment (Bosworth and Collins 2007).

With the establishment of the People's Republic in 1949, China laid the foundations for the origin of its modern innovation system. From the scientific and technological perspective, many adjustments were needed in order to break with the influence of the "Soviet system." The excessive planning, verticality, and rigidity of the system were reflected in a severe disconnect between the productive system and the scientific and technological system (Mu 2004).

It was in 2006 with the launch of the *Medium- and Long-term Plan for Science and Technology Development* (2006–2020) that China really began to reorganize its innovation system in order to achieve the integration of science and technology with the productive system.

The chief efforts were directed toward concentrating research activity into a smaller number of public institutions, orienting basic research toward applications of commercial interest, and promoting close ties with the business sector (Schaaper 2009).

In figure 4.7 we observe that it has been precisely experimental development that has seen the largest increase in R&D personnel.

At the same time, in figures 4.8 and 4.9 we observe a significant increase in R&D expenditure in both the business and higher-education sectors.

Nevertheless, these efforts to integrate the Chinese innovation system have so far not been capable of responding to the needs of regional and local industry and developing specific territorial policies favoring the creation of regional innovation systems within the national one (Lundvall et al. 2006, 295).

As a result of the system's strong centralization, there is currently a severe regional disequilibrium, such that the governments of only seven provinces contribute two-thirds of the total investment in R&D, and the

Figure 4.7. Distribution of personnel dedicated to R&D in China by type of activity (1000 FTE)

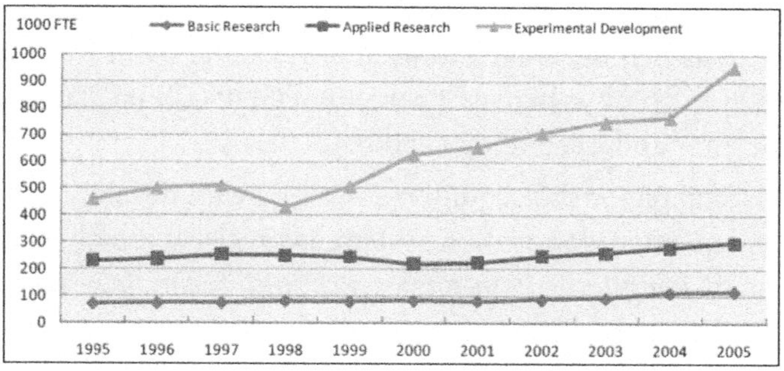

Source: NBS (2006).

Figure 4.8. R&D spending in the Chinese business sector, in billions of RMB

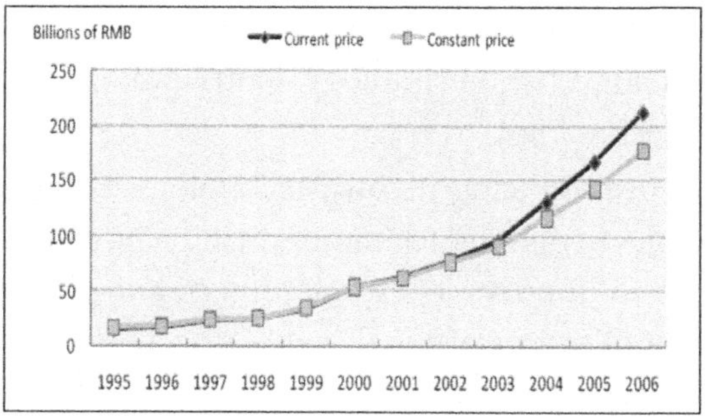

Source: OECD (2007).

Figure 4.9. R&D spending in the Chinese university sector, in billions of RMB

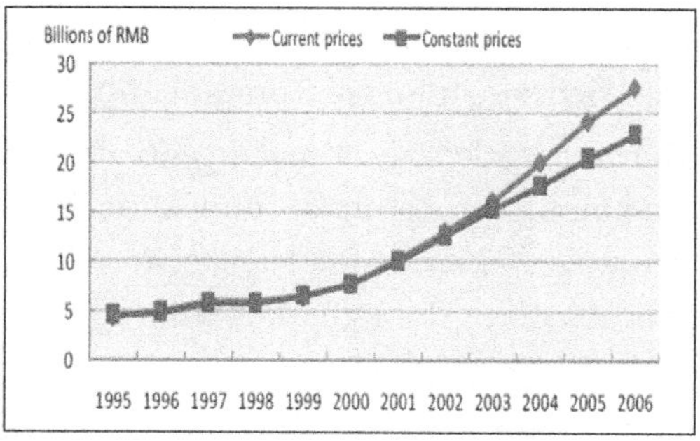

Source: OECD (2007).

remaining twenty-four account for only one-third of the total (CDTI 2010, 13). More concretely, if China dedicated 1.43 percent of its GDP to R&D expenditures in 2006, it must be taken into account that certain regions, such as Beijing, Guangdong, and Zhejiang, spent well more than this figure, even going as high as 4 percent of GDP (China Statistical Yearbook on Science and Technology 2006).

Addressing regional disequilibrium, along with developing a technologically independent innovation system, is precisely the objective of the *Medium- and Long-term Plan for Science and Technology Development* (2006–2020). As part of this plan, China has set a strategic objective of increasing R&D expenditure to around 2.5 percent of GDP by 2020. On the other hand, the plan recognizes the importance of developing scientific research, improving the research capabilities of Chinese universities, and increasing the number of qualified researchers and workers.

The National Plan is a powerful planning tool that has served to identify a series of strategic sectors: renewable energy, agricultural and food-processing technology, nanotechnology, biotechnology, and medical and pharmaceutical technology.

In addition to producing its own technology, in order to build an economy based on science and technology, China has also begun to make significant efforts to improve its exploitation of knowledge and research results. If patent applications are analyzed, China has seen a strong upward trend in recent years (see figure 4.10) and significantly, moved into fourth place worldwide in the total number of patent applications between 2002 and 2006 (see table 4.1).

This growth is also reflected in the fact that although 47 percent of patents in China are foreign-owned (Schaaper 2009, 78), if we look at figure 4.10, we see that since 2000, foreign ownership has declined significantly as resident applications have increased at a very positive rate, demonstrating China's desire to move from imitation to innovation.

In sum, China has made and is making notable efforts to achieve greater integration among the agents of innovation. Major progress has been made in encouraging knowledge transfer to the national business community and in absorbing foreign "know-how" and promoting the development of domestic technologies.

Both the private sector and the universities have become progressively more involved in market-oriented research activities, something

Figure 4.10. Patent applications, by patent office, presented by residents/nonresidents (China)

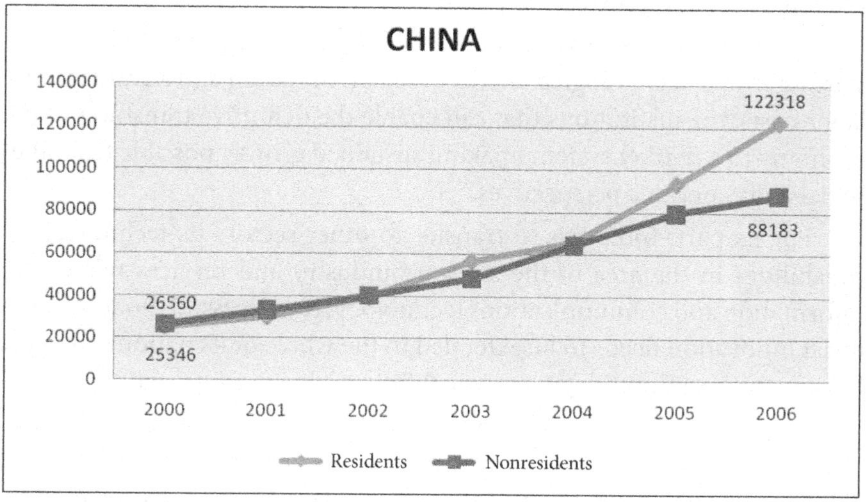

Source: WIPO (2009).

that has been reflected in significant growth in private R&D financing and in R&D work carried out in the private sector and academic institutions.

Although dependence on foreign capital and regional disequilibrium still pose significant risks at present, the National Plan represents an incredible mobilization of resources and a strong commitment to and concern for the country's future development, choosing to invest in improving its innovation system and progressing toward the latter's true independence.

Conclusions

World economic globalization and the intensification of international competition have made a change of direction evident in public policy and in the development of modern innovation systems, with a shift in emphasis from R&D efforts to technological diffusion and the search for commercial applications of new knowledge, something reflected in the growth of those global flows of knowledge that are reflected in patent indicators, as analyzed throughout this chapter.

In this regard, we have also detected the "emergence" of new technological leaders in the new global context. Thus, in the preceding pages we have sketched out some of the chief characteristics of those models of economic growth that have shown to be the most successful in Asia at present and that inevitably entail a challenge that needs to be taken into account by more industrialized countries.

Nevertheless, China is currently facing the challenge of obtaining, starting from a technological model based on imitation and advantageous labor costs, the institutions that can enable the definitive transition of its economy to a market system, making sustained growth possible from the social and economic perspectives.

For its part, India has to transfer to other sectors its technological capabilities in the area of the software industry and services related to information and communications technology. Its successful model of sectorial innovation needs to be extended to the whole of its national system by empowering the endogenous capabilities of its various regions.

It is evident that it is necessary to invest heavily in technological innovation if the aim is to compete on the international level. Economic globalization is already an indisputable reality, in which every country must make a firm and definitive choice to invest. In addition, it will no longer be possible to limit globalization to the European continent alone, as has been the case until now with the most developed countries, or to the experiences of the United States and Japan alone.

References

Aizawa, Masuo, Laxman Prasad, Moon Su-Yeon, and Zeng Guoping. 2008. Políticas de I+D en Asia: Japón, India, China y Corea del Sur. Study created for Casa Asia, Barcelona.

Asheim, Bjorn, and Lars Coenen. 2005. Knowledge bases and regional innovation systems: Comparing Nordic clusters. *Research Policy* 34: 1,173–1,191.

Bosworth, Barry, and Susan Collins. 2007 (February). *Accounting for growth: Comparing China and India.* Working Paper Series, National Bureau of Economic Research. www.nber.org/papers/w12943.

Centro para el Desarrollo Tecnólogico Industrial (CDTI). 2010. *Guía para la colaboración con China en I+D+i.* Madrid: CDTI. www.cdti.es/recursos/publicaciones/archivos/35373_1721722010113244.pdf.

China Statistical Yearbook. 2006. *China S&T Statistics Data Book 2006.* www.most.gov.cn/eng/statistics/2006/index.htm.

Coenen, Lars, and Bjorn Asheim. 2006. Constructing advantage at the northern edge. In *Regional development in the knowledge economy*, edited by Phillip Cooke and Andrea Piccoluga. London: Routledge.

Cooke, Phillip, Martin Heidenreich, and Hans-Joachim Braczyk. 2004. *Regional innovation systems: The role of governance in a globalized world*. 2nd ed. London: Routledge.

Desai, S. S. M., and Nirmal Bhalerao. 1996. *Industrial economy of India*. Mumbai: Himalaya Publishing House.

Edquist, Charles. 1997. *Systems of innovation: Technologies, institutions and organizations*. London: Pinter.

Freeman, Chris. 1987. *Technology policy and economic performance: Lessons from Japan*. London: Pinter.

Gómez Uranga, Mikel, Goio Etxebarria, and Jon Barrutia. 2009. Estudio de los cambios en los sistemas regionales de innovación a través de la evolución y diversificación de los sectores más representativos de la industria vasca. *Ekonomiaz* 70: 106–131.

Heeks, Richard. 1996. *India's software industry: State policy, liberalisation, and industrial development*. New Delhi: Sage Publications.

Laestadius, Staffan. 1998. Technology level, knowledge formation and industrial competence in paper manufacture. In *Micro foundations of economic growth*, edited by Gunnar Eliasson et al. Ann Arbor: University of Michigan Press.

Lorentzen, Anne. 2009. Las redes de conocimiento en el espacio: Reflexiones de una geógrafa sobre la literatura de los sistemas regionales de inovación. *Ekonomiaz* 70.

Lundvall, Bengt-Åke, ed. 1988. Innovation as an interactive process: From user-producer interaction to the national system of innovation. In *Technical change and economic theory*, edited by Giovanni Dosi et al. London: Pinter.

Lundvall, Bengt-Åke. 1992. *National systems of innovation: Towards a theory of innovation and interactive learning*. London: Pinter.

Lundvall, Bengt-Åke, Pataropong Intarakumnerd, and Jan Vang. 2006. *Asia's innovation system in transition*. Northampton, MA: Edward Elgar.

Mitra, Raja. 2007. *India's emergence as a global R&D centre*. Working Paper. Ostersund: Swedish Institute for Growth Policy Studies (ITPS).

NASSCOM. 2006. National Association for Software and Service Companies. New Delhi. www.nasscom.org.

Navarro, Mikel. 2009. Los sistemas regionales de innovación. *Ekonomiaz* 70: 24–59.

Nelson, Richard. 1993. *National innovation systems: A comparative analysis*. Oxford: Oxford University Press.

OECD (Organisation for Economic Co-operation and Development). 1963. *The measurement of scientific and technical activities* (Frascati Manual). Paris: Organisation for Economic Co-operation and Development.

———. 1997. "Révision des classifications des secteurs et des produits de haute technologie." Doc. OECD/GD (97) 216.

———. 2009. OECD science, technology and industry scoreboard 2009. www.oecd.org/document/10/0,3746,en_2649_33703_39493962_1_1_1_1,00.html.

Parthasarathy, Balaji. 2004. Globalizing information technology: The domestic policy context for India's software production and exports. *Iterations, An Interdisciplinary Journal of Software History*: 1–38.

Robertson, Paul, and Parimal Patel. 2007. New wine in old bottles: Technological diffusion in developed countries. *Research Policy* 6 (5): 708–721.

Sancho, Mª Rosario, and Rafael Manchón. 2006. "Éxito y perspectivas de las economías China e India: Estrategia de las empresas españolas." *Economía Industrial* 362: 39–70.

Saxenian, AnnaLee. 2002. Transnational communities and the evolution of global production networks: The cases of Taiwan, China and India. *Industry and Innovation* (Special Issue) 9 (3): 183–202.

Schaaper, Martin. 2009. Measuring China's innovation system: National specificities and international comparison. OECD Science, Technology and Industry Working Papers 2009/01, OECD Publishing. http://dx.doi.org/10.1787/227277262447.

von Hippel, Eric. 1988. *The sources of innovation*. Oxford: Oxford University Press.

5

Innovation in Services: An Application in the Case of KIBS

Goio Etxebarria and Mikel Zurbano

Translated by Jennifer Martin

Nature of Innovations in Services

Service activities have been on the margin of innovation studies, which have traditionally centered on the field of manufacturing. In spite of the growing interrelationship between service activities and manufacturing, the specific traits of this heterogeneous sector of activities with regard to industry have not favored innovation analysis development and technical change in it. *Intangibility*, information-intensive content, nonstorable character, low level of capital equipment in service, the limitation in forming economies of scale, *simultaneity* between production and consumption, *customization* of some services for certain clients, the most important *human factor* in services, the importance of a regulatory framework, and the typical smallholder and heterogeneous structure of the sector are some of the specific services traits that have contributed to services being traditionally viewed as *passive* consumers of technology or as *laggard* innovators (Howells 2001).

The growing quantitative and qualitative importance of the services sector, particularly of knowledge-intensive business services (KIBS) in Western countries within the context of the knowledge economy and new innovation-related contributions, are generating another perception concerning the combination of both aspects. This situation is promoting a broader theoretical development in the field of service innovation. The broader vision of innovation processes, which we will refer to, later creates an appropriate framework for an analysis of service innovation that is multidimensional and interactive in nature. In the growing literature on service innovation, various service innovation modes are established that refer to product, process, organization, and market innovations and stress the importance of organizational innovations insofar as they clearly contribute to improving productivity and quality of services (Sundbo 1998; Hertog 2002; OECD 2006; Howells and Tether 2004; Tether et al. 2002). Carter Bloch et al. (2008) linked the following dimensions to service innovations: innovations in the value chain and in business networks that realize the importance of interrelationships with other businesses and the importance of a regulatory framework; customer interaction and changes in customer *interfaces*; new service delivery systems; organizational changes in business structure, management, and work practices, or in the business innovation model; and the technological dimension that includes product and process innovations.

Taking this multidimensional reality into consideration directly affected the bases of service innovation analysis. In fact, besides its impact on innovation policies, the variety and importance of nontechnological innovation methods in service activities became evident. Likewise, contributions made by authors from the Groningen Growth and Development Centre group emphasized the relevance of three nontechnological dimensions (concept, interface, and delivery system) even above technological aspects in service innovation (Hertog, Broesma, and van Ark 2003). The noncentrality of technological innovation in services methods was a fundamental element in the analysis framework; this remained a differential aspect in relation to innovations in manufacturing.

However, there was a historical dichotomy between those researchers that observed service innovation as assimilable to innovation in manufacturing and those that emphasized the specificities of service innovation from a differentiation perspective (Miles 2001). Both perspectives responded differently to the matter of conditions, motives, and obstacles in service innovation. The *technologist* viewpoint that Jean Gadrey and Faïz Gallouj (1998) and John Brysson and Christine Monnoyer (2004) referred to—or *assimilationist*, called such by Rod Coombs and Ian Miles (2000) and Carter Bloch et al. (2008)—considered a concept and service innovation dynamic assimilable to those of manufacturing firms and predicted a passive view of services as recipients, not as sources of knowledge and technology. Here, preeminence of innovation continued to reside in technology. On the other hand, the *differentiating* perspective (referenced by Bloch et al. 2008) or *services-oriented* (described by Brysson and Monnoyer 2004; Tether 2005; and Rullani, Paiola, and Sebastiani 2005) stated that a specific approach must be applied to service innovation due to its distinct nature. According to this, technological innovation was only a small part of the innovations made by services companies, so innovation in services necessarily included a wide variety of nontechnological forms. This approach was even validated for the analysis of manufacturing companies in agreement with a less linear, more integral innovation concept perspective by authors like Ina Drejer (2004) and OECD (2006), and authors ascribed to the paradigm of national and regional innovation systems.

Another group of authors tried to go beyond this dichotomy and introduced a *synthesis* viewpoint (called such by Bloch et al. 2008; Sebastiani 2007; Coombs and Miles 2000) or *integrative* (Gadrey and Gallouj 1998). In fact, it is difficult to establish a clear separation between innovation in services and innovation in manufacturing because the boundary between the production of goods and services became more and more blurred, and both service and manufacturing activities increasingly "contained" operations from the other activity sector (Daniels and Bryson 2002; Howells 2004; Christensen and Drejer 2007; OECD 2006). The acceptance of this perspective had diverse implications that also impacted innovation policies because the advancement of service innovation affected all activity sectors, not just the services area.

The acknowledgment of a multidimensional character in service innovation was closely linked with the importance of the institutional environment for innovation in general, and for that of services in particular. The collaborative processes and the interaction between companies that formed part of a territory or a value chain were intimately linked to the expansion of knowledge-intensive service activities (Zurbano 2005). Yet not all service activities were equally connected to innovation environments. Furthermore, the services did not have a sufficient institutional and informal structure support for their innovative drive in comparison with other sectors. The systems for creating knowledge resources (technological centers, universities, and so on) and systems for the protection and dissemination of innovative knowledge were especially manufacturing-oriented (Miles 2001). In spite of this, large services companies had an active role as drivers of innovative activity in their provision and distribution sites. This was particularly true for knowledge-intensive services companies, as you will see further on.

Conditioning Factors and Typology of Innovation in Services

The new conception of services companies as potential innovators rather than just as recipients demanded that greater attention be paid to the variety of service innovation activities, strategies, and processes (Tether 2005). This variety was reflected in the wide range of criteria that could be used to characterize and distinguish the different types of services: capital intensity, skills and abilities, ICT usage, specialization, organizational structure, specialization, company size, and so on. In turn, these criteria were dynamic and had a direct bearing on the way in which different types of services companies innovated.

This heterogeneity and variety made the already complex process of innovation in services difficult to grasp. Various authors have made the effort to synthesize the diversity of innovation in service processes. Luc Soete and Marzella Miozzo (1989) transferred Pavitt's taxonomy for manufacturing to the services environment and summarized its innovative processes in four categories. The first corresponded to activities and *supplier-dominated* companies; they were the least innovative and their technological acquisition depended on external sources. It comprised the bulk of service activities, including public, social, and personal services. There were also *network and interactive services* that included a wide group of activities such as financial services, telecommunications, and advertising. The innovations were not so much achieved here by R&D access, but through intense interaction with clients, and were based on information and communication networks. *Scale-intensive services* with a high degree of output standardization were the third category. These were services that depended on physical networks, and therefore, on technologies developed by manufacturing firms that consequently played a very important role in their innovation specifications. Transportation and travel services, delivery, and wholesale trade fell into this category. Lastly, *scientific and technology-based* or *knowledge-intensive services* were the main creators and disseminators of new technological knowledge. They included companies and activities with close customer relationships and scientifically based ties such as engineering firms, universities, software services, laboratories, and technology centers. There were also activities that could be classified into more than just one category, as was the case with communications, which besides containing network services, was also scale-intensive and scientifically based.

Other contributions suggested new parameters in service typology such as that of adding others that best fit a greater proportion of the innovative dynamic itself to the category of *supplier-dominated services* set by Keith Pavitt (1984). This was the case of innovation at the heart of services, customer-driven innovation, paradigmatic innovation, and innovation between services (Hertog, Broersma, and van Ark 2003). The bottom line was that, in this typology, technology was not the principle instrument of the innovation in services dynamic.

Ragnhild Kvalshaugen, Katja Hydle, and Carter Bloch (2008) also established a typology that was not so technology centered. They based it on two dimensions central to service innovation for businesses in particular: the degree of standardization/customization and the degree of customer interaction. The first was important because the possibility of codifying and delivering service in a specific or independent way depended on it. The degree of customer interaction was understood as the customer involvement level in service delivery. The authors identified four types of service activities from those two dimensions. First were *expert business services* with high customer interaction and low standardization (technological services, development and design, research). *Customer business services* exhibited a high degree of standardization and customer relations. Consulting companies were classified here, for example. *Special business services* had a low interaction level and a high level of customization as in the case of engineering, legal advisory, and technical specification services. Finally, there were *standard business services* with low customer interaction (as in the case of ICT support services).

Also of interest in the study of service innovation processes was the identification of those factors that act as obstacles to innovation. Briggite Preissl (1998) noted five factors that acted as service innovation barriers and that did not differ much from those of the manufacturing sector. First were legal and bureaucratic barriers in the labor market, in fiscal legislation, in greater market regulation, and in a lack of or less protection of service innovations (difficulty with patents). Furthermore, there were financial restrictions encountered by services companies. In the third place were technical and technological-type difficulties. The limited absorptive capacity of service markets was an added barrier. The last barrier that this author cited was a lack of personnel who were qualified and experienced in innovation management, that is, a weak organizational culture of current innovation in service companies. Other types of barriers that were included in this field were linked to the nature of services: simultaneity between production and consumption, less *visibility* of the effects of innovation due to the intangible character of services, information problems, and a lower productivity growth rate (Kozlak and Stefniak 2008). Finally, the service innovation's own course was an added limitation insofar as it had a short cultural and institutional history.

Knowledge-Intensive Services in Innovation Systems

The study of innovative processes in services cannot ignore the potential of service activities to promote innovation systems. Several different studies have stressed that services, particularly *knowledge-intensive services*, particularly KIBS, had a favorable effect on innovation (Camacho and Rodríguez 2004; Hipp and Grupp 2005; Vence and Trigo 2009; Arundal et al. 2007; García Quevedo and Más Verdú 2008; Kautonen et al. 2008), especially in territories with an extensive business tapestry of small and medium enterprises (SMEs) and these KIBS established themselves as a basic pillar of national and regional innovation systems (Fischer 2001). KIBS were responsible for combining knowledge of diverse origin through active learning and through its dissemination in both manufacturing activities and the services themselves (Hipp and Grupp 2005). But they also contributed to the process of generating knowledge through social learning. This capability to generate and disseminate knowledge derived from the characteristics of the relationships between knowledge-intensive services suppliers and their customers: whether they were temporary or long-lasting relationships, from the role that KIBS played in the innovation activities of their customers, from the operations that KIBS was involved in, and from the knowledge incorporated into these services. Specifically, KIBS carried out three types of significant roles in relation to their customers' innovation capacity: They facilitated innovation by contributing to the innovative process of business clients, they were carriers involved in the transference of already existing innovations to their customers, and they were the origin or source of innovation when they played a decisive role in generating innovation in their user companies (Hertog 2002). Additionally, these services were found among those that most intensely incorporated the use of ICT, in such a way that they were leaders in adopting innovation from other sectors, and therefore played a pioneering role in the configuration and adaptation of information technology systems and software (Miles 2001).

In this way, KIBS provision generated a mutual transfer of knowledge, a process whose basis was reciprocal learning of a cumulative and incremental nature, known as the *co-production* process of innovation (Hertog 2002). This could span from momentum in achieving the merger of company-specific knowledge or from a more generic industry knowledge up to the creation of new knowledge: the environmental impact of industrial wastes, diverse software, and other advanced applications of ICT.

The companies associated with KIBS are generally characterized by a greater involvement in the active learning process and in the search for, and the accumulation of, external knowledge versus other companies more attached to "business as usual" practices that impose the re-creation of routines, concepts, and existing consultancy tools.

Emmanuel Muller and Andrea Zenker (2001) stated that the formal and informal interactions between service companies and their users not only contributed to mutually strengthening their innovative capacities but also significantly contributed to increasing the innovative potential of regional systems. The availability of these services was a very relevant aspect for increasing the innovative potential of the regions; however, market development of KIBS was restricted by a series of limitations that made proper services provision in certain territories difficult. On one hand, there is the problem of assessing the value of the services due to their intangibility. On the other hand, the provision of KIBS generates high costs, especially because of the need for highly qualified human resources. These obstacles make the provision and offering of advanced services difficult in some less dynamic countries and regions, which makes it essential to implement suitable policies to promote and boost these services.

The presence of KIBS is relatively limited in the case of the Spanish economy. Taking into account the size of sectoral employment over the total economy, it was found that while the percentage of Spanish tertiary sector employment was similar to that of the twenty-seven European Union member states (around 65 percent), the area for employment regarding KIBS is significant. In fact, the gap between the 27 percent KIBS employment rate in the case of Spain and the 32.3 percent for the EU-27 as a whole indicated a minimal degree of knowledge-intensive services in Spain (Eurostat 2008).

Although the KIBS employment growth rate in Spain had doubled that of the EU-27 during the first half of the decade (Eurostat 2008), it was not enough to reach the degree of established employment that these services held in the European Union. One aspect to consider was that the sectoral employment growth was offset by the overall growth in employment in Spain during a period of strong economic growth. However, KIBS development in Spain experienced a structural delay that cannot be disregarded when analyzing the demand side of it.

In order to observe this phenomenon and as a general approximation of the role that KIBS plays in the Spanish economy, it is interesting to study the performance of KIBS as the *origin* or *source* of innovation. To this end, using data from the Technological Innovation in Companies Survey (INE), we took the percentage of spending from four types of service functions as an indicator of the total innovation cost for innovative companies. The four functions usually rendered by KIBS were external acquisition of R&D, training, introduction of recent developments in market and design, and other preparations for production and/or delivery. In 2000, the expense for these four functions represented almost 23 percent of spending in business innovation; seven years later, this percentage remained at almost the same level because it had only raised less than one percentage point (see table 5.1). It is necessary to emphasize the unequal evolution of the different expenses linked to KIBS functions. Whereas external R&D increased almost four points throughout those years, the other three functions, particularly training and design, fell to a negative position. This loss in size was related to the positions gained by machinery, equipment, and software spending, both for arithmetical reasons and because the growing acquisition of software and equipment also incorporated training packages designed to reduce the direct expense of these concepts.

One aspect to underscore is that companies with more than 250 employees resorted to more intense spending on functions provided by KIBS in both periods. This was because the larger companies spent proportionally less on internal R&D than smaller companies, which went against the assumption that smaller size is a determinant of service outsourcing. However, the smaller ingress potential of KIBS in smaller companies may be related to the fact that service activities tend to have greater interaction with each other than with manufacturing activities. This was partially confirmed by looking at the degree in which the economic activity branches acquired KIBS-provided functions. Of the five branches that spent the

Table 5.1. Distribution of the innovation spending by Spanish businesses (percent)

	1998–2000		2007	
	>250 emp.	Total	>250 emp.	Total
Total	5,310,198	10,174,259	10,015,330	18,094,616
Distribution of total innovation costs:				
Internal R&D	46.86	50.51	36.99	39.58
External R&D	12.92	11.19	17.1	15.45
Machinery, equipment, and software	19.22	19.33	32.43	31.55
Other external knowledge	9.03	7.35	8.06	5.09
Training	2.06	2.07	0.75	0.8
Market innovations	5.08	5.27	4.18	4.64
Design, other production/ distribution	4.83	4.28	3.49	2.89
SIC	24.89	22.81	25.52	23.78

Source: INE (2009).

most on KIBS functions, three belonged to services: financial intermediation, trade and the hotel industry, and other business services (INE 2009). The trend observed was also toward a growing utilization of external KIBS on the part of smaller companies, although in a very early stage.

On the other hand, twelve of the fifteen branches that spent the most on KIBS functions had a high innovative component, and in some branches the spending on these functions was much higher than the average (23.7): This was the case in the aeronautics and space branch (41.6), other business services (36.6), pharmaceutical products (33.7), computer programs (27.7), office machinery (26.5), computer activities (26.4), and communications (26.4). In synthesis and in line with what was done a few years ago, José Antonio Camacho and Mercedes Rodríguez (2004, 193) asserted that there was a very general trend in Spain toward KIBS playing a greater role as the facilitator and knowledge-holder in industrial activities and sophisticated services from the innovative perspective. However, the role of knowledge-disseminator was somewhat weak because smaller companies did not show a change in their innovation expenditure structure in favor of functions provided by KIBS.

Conclusions

The problems of service innovation require specific handling and re-create a multidimensional and interactive paradigm of innovation where technology does not play a unique or even a central role. The majority of local/national innovations have very few possibilities of spreading to a more global level.

However, the development of ICT allows a greater number of service companies to adopt higher-quality innovations. With regard to KIBS, we find activities with great innovative potential, but with weak development in the Basque and Spanish economy, particularly concerning dissemination.

Finally, innovation policies have generally been geared toward the manufacturing sector up until now and have not covered services; or if they have, they have done so in a subsidiary manner. We believe that it is necessary to develop innovation policies explicitly aimed at service activities that promote nontechnological innovation, interact with the companies, and also focus on the side of demand.

References

Arundel, Anthony, Mina Kanerva, Adriana van Cruysen, and Hugo Hollanders. 2007. *Innovation statistics for the European service sector: Inno metrics report.* Brussels: European Commission.

Bloch, Carter, E. Adalsteinsdottir, P. O. Brehmer, J. L. Christensen, I. Drejer, K. Hydle, M.B. Jensen, J. Kuusisto, R. Kvalshaugen, and A. L. Vinding. 2008. *Service innovation in the Nordic countries: Key factors for policy design.* ServINNo final report. Aarhus, Denmark: University of Aarhus.

Bryson, John, and Christine Monnoyer. 2004. Understanding the relationship between services and innovation: The RESER review of the European service literature on innovation. *The Services Industries Journal* 24: 205–22.

Camacho, José Antonio, and Mercedes Rodríguez. 2004. España ante la economía del conocimiento: El papel de los servicios a empresas intensivos en conocimiento. *Revista Asturiana de Economía* 31: 177–201.

Christensen, Jesper Lindgaard, and Ina Drejer. 2007. *Blurring boundaries between manufacturing and services*. ServINNo project paper. Aarhus, Denmark: University of Aarhus.

Commission of the European Union. 2007. *Towards a European strategy in support of innovation in services: Challenges and key issues for future actions*. Brussels: Commission of the European Union.

Coombs, Rod, and Ian Miles. 2000. Innovation, measurement and services: The new problematique. In *Innovation systems in the service economy*, edited by Stanley Metcalfe and Ian Miles. Dordrecht: Kluwer.

Daniels, Peter, and John Bryson. 2002. Manufacturing services and servicing manufacturing: Knowledge-based cities and changing forms of production. *Urban Studies* 39: 977–991.

Drejer, Ina. 2004. Identifying innovation in surveys of services: A Schumpeterian perspective. *Research Policy* 33: 551–562.

Eurostat. 2008. *Science, technology and innovation in Europe*. Luxembourg: Commission of the European Union.

Fischer, Manfred. 2001. Innovation, knowledge creation and systems of innovation. *Annals of Regional Science* 35 (2): 199–216.

Gadrey, Jean, and Faïz Gallouj. 1998. The provider-customer interface in business and professional services. *The Services Industries Journal* 18 (2): 1–15.

García Quevedo, José, and Francisco Mas Verdú. 2008. Políticas de innovación y servicios intensivos en conocimiento. *Economía Industrial* 363: 179–186.

Hertog, Pim Den. 2002. Co-producers of innovation: On the role of knowledge-intensive business services in innovation. In *Productivity, innovation and knowledge in services*, edited by Jean Gadrey and Faïz Gallouj. Cheltenham, UK: Edward Elgar.

Hertog, Pim Den, Lawrence Broesma, and Bart van Ark. 2003. On the soft side of innovation: Services innovation and its policy implications. *De Economist* 151: 433–452.

Hipp, Christiane, and Hariolf Grupp. 2005. Innovation in the service sector: The demand for service-specific innovation measurement concepts and typologies. *Research Policy* 34: 517–535.

Howells, Jeremy. 2001. The nature of innovation in services. In *Innovation and productivity in services*, edited by Organisation for Economic

Co-operation and Development. Paris: Organisation for Economic Co-operation and Development.

——. 2004. Innovation, consumption and services: Encapsulation and the combinatorial role of services. *The Service Industries Journal* 24: 19–36.

Howells, Jeremy, and Bruce Tether. 2004. *Innovation in services: Issues at stake and trends*. Report to the Commission of the European Communities. Brussels: Commission of the European Union.

INE (Instituto Nacional de Estadística). 2009. *Encuesta sobre innovación tecnológica en las empresas 2007*. Madrid: Instituto Nacional de Estadística.

Kautonen, Mika, Marja Hyypiä, and Hanna-Mari Puuska. 2008. *Links to the international level: A study of the Finish Management-KIBS in terms of innovation system approach*. Stuttgart, Germany: XVIII RESER Conference.

Kozlak, Aleksandra, and Joanna Stefaniak. 2008. *Specificity and level of innovation in the services sector*. Stuttgart, Germany: XVIII RESER Conference.

Kvalshaugen, Ragnhild, Katja Hydle, and Carter Bloch. 2008. *Towards a taxonomy for business service innovation-literature review*. ServINNo project paper. Aarhus, Denmark: University of Aarhus.

Miles, Ian. 2001. Services innovation: A reconfiguration of innovation studies. PREST Discussion Paper Series. Manchester, UK: University of Manchester. http://les.man.ac.uk/PREST/.

Muller, Emmanuel, and Andrea Zenker. 2001. Business services as actors of knowledge transformation: The role of KIBS in regional and national innovation systems. *Research Policy* 30: 1,501–1,516.

——. 2006. *KISA—Innovation and knowledge-intensive service activities*. Paris: Organisation for Economic Co-operation and Development.

Pavitt, Keith. 1984. Sectoral pattern of technical change: Towards a taxonomy and theory. *Research Policy* 13: 343–373.

Preissl, Briggite. 1998. Barriers to innovation in services. SI4S Topical Paper No. 2. Oslo, Norway: STEP Group.

Rullani, Enzo, Paolo Barbieri, Marco Paiola, and Roberta Sebastiani. 2005. *Intelligenza terziaria motore dell´economia*. Milan: Franco Angeli.

Sebastiani, Roberta. 2007. La traiettorie dell´innovazione nei servizi: Esperienze e confronto. www.unicatt.it/economia.

Soete, Luc, and Marzella Miozzo. 1989. Trade and development in services: A technological perspective. Working Paper No. 89/031. Maastricht, Netherlands: MERIT.

Sundbo, John. 1998. *The organisation of innovations in services.* Roskilde, Denmark: Roskilde University Press.

Tether, Bruce. 2005. Do services innovate (differently)? Insights from the European innobarometer survey. *Industry & Innovation*, 12 (2): 153–184.

Tether, Bruce, Ian Miles, Knut Blind, Christiane Hipp, Nicola de Liso, and Giulio Cainelli. 2002. Innovation in the service sector: Analysis of data collected under the community innovation survey (CIS2). CRIC Working Paper 11. Manchester, UK: University of Manchester.

Vence, Xavier, and Alexandre Trigo. 2009. Diversity of innovation patterns in services. *The Service Industries Journal* 29 (12): 1,635–1,657.

Zurbano, Mikel. 2005. "Services, networks and territory: The case of MCC in the Basque Country." *The Service Industries Journal* 25 (4): 547–563.

6

Corporate Strategies, Competition, and Technological Innovation among the Operators of Maritime Transport Services

Iñaki Alcedo Momoitio and José Moreno Díaz

Translated by Kristin Addis

The maritime transport services made available to users by longshoremen and freighters have special characteristics that distinguish them markedly from other types of services. Transport services have no material form that can be seen, nor is it possible to measure them by empirical methods or systems to determine their quality, at least not before using them; they are not tangible. This leads to their evaluation after the fact being subjective, based on the unique experience of each user. The lack of tangible physical evidence increases the level of uncertainty of a new client, but furthermore, one innate characteristic of transport services is that they cannot be stored for later sale or for times of high demand. Additionally, maritime transport understood as a service offered to a market of potential users is never the same from one time to the next, given that the physical voyage of the goods regularly undergoes changes that in some cases may be significant.

Despite this, there is a phenomenon, primarily in the industry of regular container lines, in which the users of the service see and understand transport as a consumer product. It is a trivialization of a complicated process (*commoditization*) that is manifest in a service that is now considered to be a simple object for economic transaction and its purchase and sale, based almost exclusively on price as a determining factor.

In this context, new technologies have made it possible to introduce innovations in products, services, and business processes, allowing companies to distinguish themselves from their competition, increase their productivity, and decrease their costs. Because of this, it is necessary to adopt these technologies in order to improve aspects of competitiveness like economies of scale, quality, and service, which are so important in all sectors. The purpose of this chapter is to analyze the transformations that have taken place in businesses in the sector of operators of maritime transport from the point of view of innovation. Our analysis presents evidence concerning the uses and applications of technological improvement in the activities of these companies and their relationship with the processes of innovation and quality, the level of productivity, and the development of new competitive strategies. We also discuss compatibility among traditional competitive strategies based on costs, and new strategies based on innovation and quality, as elements of constant improvement in a company.

The maritime industry, which develops the means and systems necessary to facilitate the movement of goods among the different countries of the globe, has undergone significant changes in recent decades. The maritime sphere is experiencing the appearance of emerging countries that require and export large quantities of goods and encourage the appearance of new routes parallel to the existing traditional ones between developing countries and industrial areas.

For the operators of the maritime transport industry, the turnabout in the world economic system with the present crisis has meant a combination of challenges to face imaginatively and innovatively, keeping in mind that it is a question of surviving and growing or disappearing. Specialization is one of the directions taken by many of the more dynamic and creative companies, though of course there is no single solution, given that such specialization requires enterprising spirit, proactivity and a certain amount of courage. The ship-owner companies in this sector have also ventured in recent decades into the application of economies of scale, which has translated into significant increases in the sizes of the ships, and on their involvement in harbor operations through direct participation in stevedore companies, warehousing, and comprehensive logistics.

Using similar technologies for the equipping of ships and similar computer systems based on intranets or on the public Internet, with some loading and unloading systems in the ports being completely identical and shaped by ever-stricter government antimonopoly policies, the operators of regular container lines have only a meager margin at their disposal for differentiation among themselves. In this context of extreme intensity in the battle for a competitive advantage, creativity transformed into innovative vitality is considered an absolute necessity for the development of ship-owner companies. It is here, in this context of uncertainty, where innovation driven by the company that creates and gives expression to a new product or service in the market will provide a certain competitive advantage that in many subsectors may be essential for successful survival in the purest *Schumpeterian* style (Schumpeter 1934, 1942).

Innovation in Maritime Container Services

Innovative creativity in the maritime world is found in two general areas that involve a certain complexity: the physical means of production and management systems. On the one hand, the tremendous development of the technology used in naval architecture has produced important innovative elements that ship-owner companies have applied to their ships. Similarly, the development of containers (new materials, different designs), new loading and unloading systems in the ports, and warehouse design offer a rich casuistry of novel elements. It is a case of technological innovation applied to the means of production in this market.

Business administration and organization have also undergone profound transformations. Thus, the internal structure of the companies has evolved from the traditional centrality of established ship owners toward a more horizontal, decentralized, representative, and participatory design. With respect to the financing of these companies, the influx of capitals that are totally foreign to the maritime world (and therefore unfamiliar with its nature and the inherent risks involved) has been another constant feature for only three decades. Finally, the strategic policies of expansion and company growth have changed in the same direction mentioned previously—that is, from projects for organic internal progress using one's own means to gradually beginning to aim at growth using external means, either through foreign financing or through policies of purchase, merger, or association with greater or lesser organic integration (horizontal and vertical expansion).

The present perception of the maritime industry with respect to innovation is that the ability to gain a competitive advantage through technological advances and their applications is rather limited because the availability of such enterprising weapons is absolute for all competitors (with certain exceptions), and furthermore, their application with beneficial effects on the ability to lead the market are scant and limited in time. It seems clear, nevertheless, that innovative models for management are truly the differentiating elements. One would find oneself, therefore, in a competitive environment not only in products and services (to some extent) but also in models of business administration. Innovation in the context of maritime business in general and specifically in the context of container transport is driven by the intense pressure exerted by competition, which continually requires the creation of new forms, methods, and strategies to obtain a position of competitive advantage, however ephemeral it may be.

Three clearly innovative though generic aspects should be examined, however, when describing a market characterized by low profitability and extreme competition: productive specialization, differentiation of markets (Matthews 2007; Porter 2010) and new models of business administration.

Traditional Innovation: Specialization

Little by little, the traditional regular-line ships that used to transport cargo in heterogeneous packages, the so-called cargo liners, are being progressively replaced by ships that are more sophisticated in their design and architecture. This means that cargos of various types of goods, principally bulk cargos of liquids and solids, will embark exclusively on ships designed specifically for them.

This specialization will continue to be emphasized until it reaches the most exclusive sophistication possible at present with units designed solely for specific types of transport. This technological transformation became particularly prominent with the appearance in 1959 of containers, which brought about the design of ships purely for containers beginning in 1960; the development of such ships has not yet come to a halt. Stemming from the original and classical innovative proposal (the specialization of the product), an expansion in the same direction has been seen in the maritime market, an expansion that affects not only the technical aspects of the ships but more profoundly the very concept of the business.

Thus, in recent years, companies with centralized management have been giving way to those administrated by units of a clearly horizontal nature. In view of the general context of internationalization, the relaxation of administrative ties between countries and the considerable importance of financing in business ventures, a deeper specialization has been designed, focusing on different alternatives that cannot be integrated, such as the acquisition and ownership of ships (maritime leasing companies), their commercial use (large container operators), their operative administration (companies that bring together the administration of a large number of ships belonging to different owners), the development of financial products designed for this industry, and even innovative technological design. The challenge is to perfectly define the fundamental and central activity of the company, subcontracting the administration of the remaining activities initially included in the group to other companies in the same corporation or outside of it, which are specialists in such duties in their own right.

The entire process is easier today due to certain instruments available to the majority of companies (*enablers*), such as technologies having to do with information; investment facilities in different countries; the existence of attractive fiscal packages for the location of companies; the variety of investors available to take on new projects; the interculturality of businesses, which makes it possible to participate simultaneously in multiple companies of enormous diversity; administrative techniques for comparison among companies (*benchmarking*); and the demand for transparency in corporate administration, which facilitates the transmission of knowledge.

Innovation Based on Differentiation

Differentiation is a concept closely linked to innovation because the latter is understood, in the present context, as a process for improving the monopolistic and competitive power of a company through the creation of something new and different.

Traditionally, in the maritime sector the markets characterized by a low degree of differentiation among ship owners are those oriented around the product to transport, the so-called "goods markets" (bulk carriers, oil tankers, and chemical products tankers). On the other hand, the sectors that show high degrees of differentiation among ship owners focus on markets that are highly specialized or novel, or that have high requirements of security, operation, or type of work (car carriers, cruise ships, open-hatch bulk carriers, emergency support vessels for oil platforms, special tugboats, suppliers and pontoons for the distillation and warehousing of crude oil).

Specifically, the homogeneity mentioned here leads to great invisibility among the participating companies. The ships are very similar, the containers they use must necessarily be of strictly equal measurements no matter the operator, and the terminal ports and trucks used for deliveries do not differ from each other. This situation requires considerable effort on the part of ship owners to show themselves to be different from the competition by providing evidence that their service is better than that of the others. The aspects that operators normally stress for this differentiation vary and are often interrelated: innovation in quality, geographical coverage, image in the market, and differentiation by price.

Innovation in Quality

Taking as a point of departure the fact that the routes between fixed ports are stable with respect to time and the duration of the voyage, the ship owners insist in their marketing campaigns on various concepts that make quality stand out: reliability, frequency of departures, flexibility, and the complete availability of ships and equipment (containers) to the users. With respect to reliability, an increase in sensitivity to matters of security and the environment is seen in some sectors more than in others.

Logically, in the sector of raw energy materials (oil, gas, and chemical tankers), the degree of evolution toward quality with respect to the environment is greater than in container transport. International regulations (the International Ship and Port Facility Security Code and the International Convention for the Safety of Life at Sea, among others) created by the International Maritime Organization (a dependent of the United Nations) have clearly promoted this change of criteria (Lagoudis, Lalwani, and Naim 2006; Lagoudis and Theotokas 2007).

Furthermore, some companies have reinforced the role of the Technical Instructions (TI), offering from their Internet pages documentary systems that are very simple for the user in order to facilitate reservations, billing, tracking of shipments and the like. Even the concept of innovation is waved high as a flag of quality and is included in publicity campaigns, on the web pages of shipping companies and in advertising in specialized journals.

Geographical Coverage

A company achieves a certain distinction from others when it extends its geographical coverage of ports or when it provides specialization in very specific markets in which it has positioned itself and maintains a recognized record of quality. Interestingly, in the past decade, there has been innovative movement toward local markets in an attempt to penetrate also into small and geographically limited submarkets that previously were of no interest to the large operators. The need to increase the levels of the market drives this desire to attain specific niches that previously were not of sufficient importance.

Image in the Market

This is an aspect that is very much tied to the business marketing of services, although in this market, where the visibility of the firm is limited, the initiative to create a market image is not easy to find. In recent years, the large corporations have tried to win various prizes awarded by publications in the sector (Containerization, Lloyd's List) in spheres such as respect for the environment and emissions reduction policies, excellence in transit times (punctuality and reliability), and sophistication in the web services they offer to their clients. Corporate policy with respect to the environment has rapidly gained ground in the policies of the large shipping companies who underline in all possible forms of marketing that their sensitivity to these themes is of the utmost importance.

Differentiation by Price

This is a standard option very much used by the operators of regular lines and that gives rise to important consequences, not all of which are satisfactory for either the individual actors or the market as a whole. The option of lowering carriage fees in theory attracts a great number of clients away from other competitors. Nevertheless, given the almost identical equality of the services, a later lowering of price on the part of competing ship owners undoubtedly has the same effect. In the end, systematic reductions in carriage fees aimed at maintaining the levels of occupation of the largest ships have contributed to a general depreciation of tariffs in many cases to unsustainable limits, generating a sequence of events that reduces operational margins more and more. The innovative virtue, in any case, is limited in this area of competition.

Innovation in Administration Processes

Innovation that is not strictly technological in physical terms is based on the development of new concepts for the administration of the business. The regular container line has been especially active in this type of renewal, out of the need to respond to changes in the market and as an innate attitude toward the proactivity of a sector that is extremely dynamic. The traditional maritime industry was characterized by a business structure based on the ownership of the ships and their use through management that was generally direct. The new need for considerable capital to continue operating highly specialized ships, a more professional administration, and the explosion in the financial market of new business structures determined to invest but also to control their investments have brought about a gradual shift from traditional methods of business government to the present more professional methods, on a larger scale and with more global coverage.

From this new culture of business administration that makes use of private funds from various sources that are not always maritime, there have arisen more independent business policies that use more sophisticated instruments of strategic planning than in the past. In this sense, among the innovative policies that have been developed in recent decades, those related to the following fields of the business dynamic stand out: mergers and acquisitions, separation of ownership and the use of the means of production, innovative policies in the relationships between client and provider, partnerships, expansion to related sectors, and ship owners involved in the new strategy of logistics: operational innovation.

Mergers and Acquisitions

In the market of the regular lines, the need to achieve satisfactory results for the stockholders has put great pressure on the search for innovative alternatives. One of the paths used extensively in the last two decades by the most important operators has been the strategy of the merger with or purchase of competitors holding a level of the market in some cases greater even than that of the developers themselves.

The practical reasons that drive these tendencies are many. On the one hand, world coverage facilitates greater power of negotiation with providers of all types: oil companies for the fuel for the ships, terminals, and even port authorities in the case of loading and unloading and the makers of containers. On the other hand, the availability of large fleets offers flexibility in scheduling departures, setting itineraries, and substituting ships for other ships according to the swings of demand or in the interests of the company. Another reason is that industrial consolidation establishes barriers to entry that are more difficult to cross for competitors who might be interested in entering the market.

Separation of Ownership and the Use of the Means of Production

The general atmosphere of the maritime market today is basically favorable to the development of companies focused on very specific compartmentalized activities, even though they may be highly sophisticated companies with a common headquarters. The evolution of the concept of company from the traditional and family-run to the present is supported by various factors such as the following:

- *Dynamism in administration*: Instead of the long-term vision of the traditional ship owner, present-day operators can enter and leave markets with much greater ease due to the flexible structure of design companies and to the lesser need for long-term financing.
- *Subdivision of activities in output centers*: Until recently it was estimated that control of the different phases of a transport process within a single organization would produce complementary benefits when presenting a corporate image. Nevertheless, the present-day access to specialized providers is at the disposal of the clients, and therefore, the absolute vertical control of related activities is not believed to contribute any special added value. Thus, the primary companies have completely differentiated their division of logistics from port terminals or gas tanker ports.
- *Specialized administration*: Far from the former division of the different submarkets, with modern agencies, with their concentration on specific, limited and specialized activities, the ship-owner company gains the maximum knowledge of its field, establishes solid relationships with the clients, and creates an atmosphere favorable to innovation.
- *Financing replaced by contracts of use*: Regardless of the size of the operator, operators prefer to avoid direct financing by various means (stocks, public offers, issuing bonds, and so on) and sign contracts of use of the ships once they are built by financial companies that support such an initial process.

Innovative Policies in the Relationship Between Client and Provider

The negotiating power between client and provider is swinging significantly toward the consumer such that the providing companies have had to adjust their policies toward a greater understanding of their clients. As a priority, ship owners have had to contract agencies that had broad approaches and flexibility in their ideas, and that kept themselves open to what the client truly wanted. This rapprochement to the client would have had to be eclectic, manageable, and innovative; it would not have let itself be led exclusively by the market, but rather would have generated the policies of leadership that influenced it (Luethje and Herstatt 2004).

Partnerships

Partnerships are a good method of corporate expansion because they offer the operators various tools that in many cases are complementary to those mentioned previously, and that allow less burdensome commercial and operative actions to be taken to achieve their objectives. It should be noted that a high-quality transport service does not necessarily require that the means of production be the property of a single organization; rather, it has been shown that collaboration among companies entails, under normal conditions, financial advantages and maximization of use of all sorts of resources (Lei et al. 2008). The benefits of this collaboration will be in the excellence of the administration of those resources in a coordinated and efficient way for the purpose of offering the client a real and interesting added value. Thus, at present, the twenty primary world operators are integrated into three large partnerships: the New World Alliance, formed by APL, MOL and Hyundai; the Grand Alliance, with P&O Nedlloyd, NYK, Hapag Lloyd, OOCL and MISC; and in third place, the so-called CKYHS Alliance with Cosco, K-Line, Yang Ming Line, and Hanjin-Senator (Barry Rogliano Salles Shipbrokers 2009).

Expansion into Related Sectors: Ship Owners Under the New Logistics Strategy

The various operators have created strategies aimed at specific objectives such as operational efficiency founded basically on a greater utilization of the means of production—that is, the ships. This is translated into many different goals: an intensive search for an increase in cargo so that ships always carry the maximum load; a greater frequency of trips to increase the time spent on productive use; and therefore, a very important effort to reduce time spent unproductively (ship breakdowns, time spent waiting in port) and an increase in the effectiveness and speed of port operations. This has brought about a movement toward total logistics.

A second important step toward total logistics came to fruition in a Japanese operator, Nippon Yussen Kaisha (NYK). In the country that invented the philosophy of Just In Time (JIT), various and important clients of this ship owner sought distribution and warehousing services. NYK has consolidated its presence more and more in the distribution and warehousing market since the late 1980s and in the 1990s and has become a big competitor in the market of total logistics, in which there are real specialists like Exel or Tibbet and Britten. Others have followed in their footsteps, such as P&O Nedlloyd Logistics, which in 2002 bought the United States company Gilbert in order to be able to offer new services in that country and also provide a network of distribution to international clients. Similarly, APL Logistics acquired the Gatx group, dedicated to warehousing and also based in the United States.

Operational Innovation

Starting from the most innovative operational aspect and supported by updated methods of company management and administration, operators have been developing enterprising policies and initiatives that should also be emphasized. Radical innovation in operations—and not just continual improvement—can separate a company from its competitors and shake up a whole sector. In an economy in which the client rules like never before, operational innovation offers a significant and sustainable way to get ahead of the rest and to maintain that advantage. This operational innovation should not be confused with operational improvement or with operational excellence. These terms refer to reaching high performance through already existing modes of operation; that is, ensuring that the work is done as it should be for the purpose of reducing errors, costs and delays, but without changing fundamentally the way in which this work is carried out. Operational innovation involves creating totally new ways of taking orders, developing products, providing service to the clients, and carrying out other company activities. Operational innovation is a qualitative leap: it takes the company to a whole new level, allowing it to focus on the creation of additional changes—refinements of innovation—that keep it ahead of its competitors until the inevitable moment when a new wave of innovation is required.

Under this strategy, maritime transport operators have developed different mechanisms, not always clear, to adapt quickly to the demands of the market: increasing the size of the ships, streamlining existing routes and slowing the speed of service, positioning the ships for maximum operational capacity, general restructuring of services, and efficient management of the movements of containers.

Conclusions

The maritime industry has always been distinguished for its innovative and enterprising character, perhaps as a natural result of its inherently international character or its geographical ubiquity. In spite of the fact that this market is essentially global and that, therefore, it requires global structures of organization, one path for innovation is the search for specialized local markets. Another path for innovation in this industry is the constant search for differentiation from other competitors. Since even before the term *benchmarking* was coined, ship owners and operators in the maritime market already used unconsciously the techniques of continual improvement by comparison. But an analysis of the innovative process in the world of ocean transport sometimes reveals a sector that seems to promote technology excessively, leaving creativity in administration and methods in the background.

All in all, the creation of clear policies aimed at developing creativity and innovation in these companies is absolutely fundamental for the enterprising process to flow in a regular and effective manner, with general participation. Company management, in moving toward these objectives, must be specific about criteria; methods; and simple, objective and unambiguously defined channels so that the participation of all involved parties can arise in the most natural way.

References

Barry Rogliano Salles Shipbrokers. 2009. *Shipping and shipbuilding markets in 2008*. Paris: Barry Rogliano Salles, 1–9.

Lagoudis, Ioannis N., Chandra S. Lalwani, and Mohamed M. Naim. 2006. Ranking of factors contributing to higher performance in the ocean transportation industry: A multi-attribute utility theory approach. *Maritime Policy and Management* 33 (4): 345–369.

Lagoudis, Ioannis N., and Ioannis Theotokas. 2007. The competitive advantage in the Greek shipping industry. *Research in Transportation Economics* 21: 95–120.

Lei, Lei, Chunxing Fan, Maria Boile, and Sotiris Theofanis. 2008. Collaborative vs. non-collaborative container-vessel scheduling. *Transportation Research Part E: Logistics and Transportation Review* 44 (3): 504–520.

Luethje, Christian, and Cornelius Herstatt. 2004. The lead user method: an outline of empirical findings and issues for future research. *Research and Development Management* 34 (5): 553–568.

Matthews, Steve. 2007 (July 11). Liner profits under pressure. *Lloyd's Shipping Economist 2007* 6: 15–27.

Porter, Janet. 2010 (December 6). Drewry warns box sector still fragile. *Lloyd's Shipping Economist 2010*: 65–74.

Schumpeter, Joseph. 1934. *The theory of economic development: An inquiry into profits, capital, credit, interest and the business cycle.* Translated from German by Redvers Opie. Cambridge, MA: Harvard University Press.

———. 1942. *Capitalism, socialism and democracy.* 2nd ed. New York and London: Harper and Brothers.

7

Innovative Management and Changes in Organizational Culture at the University of the Basque Country

Francisco Panera Mendieta and Sabino Ayestarán Etxeberria

Translated by Jennifer R. Ottman

The University of the Basque Country (Universidad del País Vasco–Euskal Herriko Unibertsitatea: UPV/EHU) is a public university system organized in three campuses of different sizes (Araba, Gipuzkoa, and Bizkaia) corresponding to the political and administrative divisions of the Comunidad Autónoma del País Vasco/Euskal Autonomia Erkidegoa (CAPV/EAE, the Autonomous Community of the Basque Country).

With around 45,000 students, almost 4,800 professors, and more than 1,600 administrative and service staff, the university has 108 departments and thirty-one centers (seven in Araba, eleven in Gipuzkoa, and thirteen in Bizkaia), plus three other affiliated centers.

The university is organized in a complex cellular structure in which collegial bodies and individual posts are divided territorially, separated by functions and competencies that vary widely by center and department. At the top of the hierarchy is the rector's office, organized by function and territory, and the administration with its centralized services. Finally, all the collegial bodies include weighted representation of the three university constituencies: students, professors (with and without administrative responsibilities), and administrative and service staff.

For some time, the UPV/EHU has been very aware of the political, economic, and social challenges it must inevitably face. These changes affect the institution's very essence: its distinctive political, economic, and social mission. The university is no longer only a free space for intellectual reflection, critical thinking, and the pursuit of scholarly knowledge but is also required to train professionals capable of satisfying the needs of the productive system and effective researchers who obtain useful and applicable results for society.

At the same time, European harmonization and/or convergence manifests itself in ensuring the equivalence and interchangeability of these services and their results, to the same extent that this is becoming the case in the productive sectors, across the entire European Higher Education Area. In other words, the aim is that the training, competence, and qualifications of any European university graduate be independent of the country or the university where he or she was trained.

This chapter describes the stages through which our university has passed along this path toward management innovation, one that has entailed significant changes in organizational culture. This has been a lengthy process that has stretched from training programs in quality management and encouraging improvement teams to quality-assurance certifications (ISO standards) and Basque Foundation for Quality (EUSKALIT) quality-management awards (European Foundation for Quality Management [EFQM] Excellence Model).

At present, we find ourselves in a phase of generalizing the implementation of quality-assurance systems in our centers, at the same time that numerous centers have received external recognition through the Silver Q Award for implementing the EFQM Excellence Model.

Model of Organizational Culture

The "competing values" model put forward by Cameron and Quinn (figure 7.1) includes two bipolar dimensions. The first dimension represents the organization's internal or external orientation. In the first case, the organization is focused on its own affairs, on its internal processes, and/or on its personnel. In the case of an external orientation, the organization is fundamentally concerned with its relations with its external environment.

The second dimension involves orientation toward flexibility or toward control in relation to the form in which authority is exercised within the organization.

By combining these two dimensions, four types of cultural orientation are obtained: clan, hierarchy, market, and adhocracy.

A particular organization may be strongly marked by one orientation, by several, or by none. In this last case, the organization lacks a clearly defined culture.

Figure 7.1. Cameron and Quinn's "competing values" model

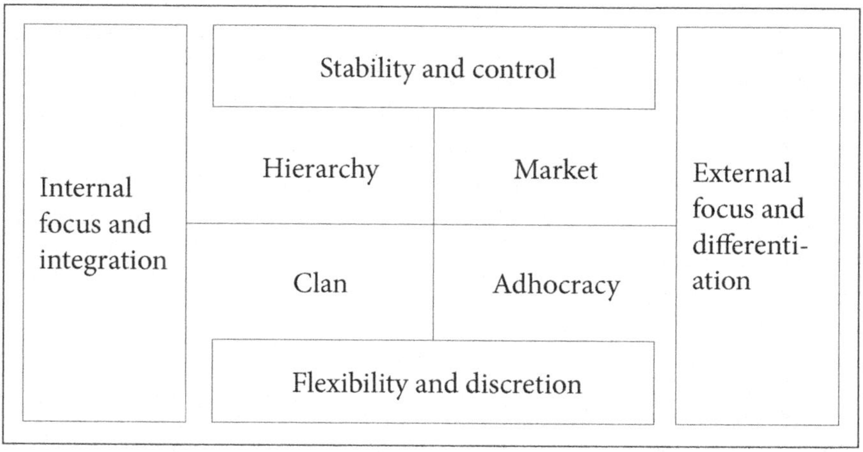

Source: Cameron and Quinn (1999).

These dimensions also produce quadrants in which values contradict one another or "compete" along the diagonals. The upper-right quadrant, for example, is associated with values that emphasize an internal and controlling orientation, whereas the lower-left quadrant is associated with values that emphasize an external and organic orientation. Similarly, the upper-left quadrant is associated with values that emphasize an external and controlling orientation, whereas the lower-right quadrant emphasizes internal and organic values. These competing or opposed values give the model its name: the *Competing Values Framework*.

The strength of the dimensions and the richness of the resulting quadrants allow us to identify each quadrant as a *culture type*. Each represents basic assumptions, orientations, and values: the same elements that comprise an organizational culture.

The Four Culture Types

In the following sections we will discuss the four culture types: hierarchy culture, market culture, clan culture, and adhocracy culture.

Hierarchy Culture

Hierarchy represented an ideal organization until the 1960s because it enabled the provision of stable, efficient, and highly consistent products and services. In a relatively stable environment, tasks and functions could be integrated and coordinated, uniformity of products and services was maintained, and both workers and their jobs were controlled. Clear chains of command and decision-making, standardized rules and procedures, control, and accountability mechanisms were valued as the keys to success.

The organizational culture compatible with this form of organization is characterized by a *formalized and structured* working environment. Procedures govern what people do. Effective leaders are good coordinators and organizers. Maintaining smooth functioning of the organization is important. The organization's long-term goals are stability, predictability, and efficiency. Formal rules and policies are what make the organization work.

Market Culture

In contrast to a hierarchy culture in which internal control is maintained through rules, specific tasks, and centralized decisions, a market culture operates chiefly through market economic mechanisms. The emphasis is placed on *managing transactions* (exchange, sales, contracts) with external agents for the purpose of creating competitive advantage.

The central values that dominate this kind of organizational culture are *competitiveness* and *productivity*, both of which are achieved through a strong emphasis on effective positioning and external control.

A market culture is a *results-oriented* working environment. Leaders are fierce, competitive, and demanding producers. The long-term goals are competitive activities and the achievement of ambitious goals and objectives. Success is defined in terms of good market positioning and penetration. It is important to beat the competition and be a market leader.

Clan Culture

A *clan* kind of organization resembles an extended family more than an economic entity. Instead of the rules and procedures of hierarchical organizations or the central competitive benefits of market organizations, the typical characteristics of clan-type enterprises are work teams, employee investment in programs, and corporate commitment to the employees.

Clan culture is characterized by a friendly working environment in which people share a great deal of themselves. It is like an extended family. Leaders think of themselves as mentors and are perhaps seen as paternal figures. The organization works on the basis of loyalty and tradition. The existing commitment is high, and the organization's long-term goal is fostering individual development with a high level of morale and cohesion. Success is defined in terms of internal climate and concern for individuals. The organization reinforces *teamwork, participation,* and *consensus.*

Adhocracy Culture

The roots of *adhocracy* culture are the "ad hoc," referring to contingent, specialized, and dynamic entities. The primary objective is to promote adaptability, flexibility, and creativity.

In contrast with hierarchy and market cultures, this culture type does not have centralized power or well-established relations of authority or chains of command. Here, power flows from individual to individual or from workgroup to workgroup, depending on the problem being addressed at any given moment. There is a strong emphasis on individuality, on taking risks, and on looking toward the future.

Normally, units with an adhocracy culture are created within larger organizations that have a different dominant culture. Many problems can arise from this.

In effect, adhocracy culture is characterized by a dynamic, entrepreneurial, and creative working environment. Individuals seek to stand out and take risks. Effective leadership is visionary, innovative, and risk-oriented. Preparation for change and the constant posing of new challenges are important elements.

Stage One: Developing, Diffusing, and Extending the Culture of Quality

The first driver of change in modeling a culture of quality and quality management is training. Numerous courses were and are offered at the UPV/EHU for the purpose of developing quality-assurance programs. These courses have included principles of quality management; promoters, leaders/coordinators, and facilitators of improvement teams; the ISO

9001/2000 standard and internal auditing applied to university teaching; training of external assessors and auditors (Agencia Nacional de Evaluación de la Calidad y Acreditación: ANECA [National Agency for Quality Assessment and Accreditation]); training in the design and implementation of process management and of an Internal Quality-Assurance System at a university center; strategic planning, process mapping, and management indicators; teaching teams and leadership in the classroom; summer courses in Donostia–San Sebastián on improvement teams, academic performance, and applying the EFQM quality-management model in university centers; management skills for management teams and application of the EFQM model; training and refresher courses for external assessors in the EFQM model (EUSKALIT Assessors Club); internal and external communication; and the "5S" methodology for workplace organization and improvement (secretariat, administration, archives, student services, teaching and research laboratories).

Along the same lines, center administrators were encouraged and supported in holding special events (Quality Days) at each center to celebrate and recognize the work of all those who participated in these programs, as well as to publicize their achievements. Publicizing actions taken and success achieved is a way of extending the culture of quality, and this was a fundamental and high-priority objective, even if its effects might only be evident in the very long term.

At the same time, we were very clear about the fact that training without action loses its significance as useful learning and often becomes no more than sterile "book learning."

For that reason, our resources, scarce as always, were dedicated to offering programs that combined training and action. This was training that was no longer based simply on learning, but rather on immediately applying what was learned. Thus, we had courses for center administrators on strategic planning, process management, teamwork, the EFQM model, "5S," and other topics, in which the objective was to write a strategic plan, develop a process map, work in a team, carry out a self-assessment and implementation, and so on.

In sum, extending the culture of quality was always based on the combination of training and action, was promoted by publicizing actions and achievements, and constantly revolved around one basic concept: *customer service, employee satisfaction, and continuous improvement.*

During this first stage, change was also promoted through the TQM principles of *employee participation and process improvement,* by encouraging teamwork and fostering "improvement teams."

The promotion of improvement teams took place through training courses, as well as through encouraging and following up on their work, with the aim of achieving improvements in processes and results. These activities were coordinated by the quality-management offices at the three campuses, with support from the professorial chair in quality management.

Stage Two: The Network of Centers of Innovative Management (Red de Centros de Gestión Innovadora: RCGI) and the Assessment and Certification Processes

The second major stage in this process of management innovation was linked to the establishment in 2001 of the Network of Centers of Innovative Management (Red de Centros de Gestión Innovadora: RCGI).

The RCGI was established in March 2001 and consisted of five centers in its first incarnation: the Bilbao Higher Technical School of Engineering (E.T.S. de Ingeniería-Ingeniaritza G.E.T.) and the Portugalete Higher Technical School of Nautical Studies (E.T.S. de Náutica y Máquinas Navales-Nautika eta Itsasontzi-makineria G.E.T.) for the Bizkaia campus; the Donostia–San Sebastián School of Technical Industrial Engineering (E.U. Ingeniería Técnica Industrial-Industri Ingeniaritza Teknikorako U.E.) and the Donostia–San Sebastián Nursing School (E.U. de Enfermería-Erizaintzako U.E.) for the Gipuzkoa campus; and the Vitoria-Gasteiz School of Technical Industrial Engineering and Topography (E.U. Ingeniería Técnica Industrial y Topografía-Industria eta Topografia Ingeniaritza Teknikoa U.E.) for the Araba campus.

The network's fundamental objective was to share the path toward and the efforts to achieve quality management, and more specifically, to work toward excellence together, by sharing each member's best practices and learning from one another in a spirit of cooperation.

The reference model chosen for adaption to the UPV/EHU was the EFQM Excellence Model in quality management, a model also adopted by the Basque government, which has strongly promoted it in the CCAPV/EAE through the Basque Foundation for Quality (Fundación Vasca para la Excelencia-Bikaintasunerako Euskal Fundazioa: EUSKALIT).

The implementation of ISO standards in all areas of management was also chosen as a set of best practices for reference, and in this first version of the network, the Donostia–San Sebastián Nursing School, obtained ISO certification for its training processes; the Donostia–San Sebastián School of Technical Industrial Engineering did the same for some of its management processes.

Each of the network centers was visited in turn, following the same methodology, thereby providing an opportunity to learn about the different realities and very diverse problems faced by the university's widely varied disciplines: engineering, the experimental sciences, health sciences, social sciences, and humanities.

In the RCGI's second edition, in the 2002/2003 academic year, the Vitoria-Gasteiz Nursing School (E.U. de Enfermería-Erizaintzako U.E.), an affiliated center, was added (Araba campus). Also noteworthy was the success achieved by the Donostia–San Sebastián School of Technical Industrial Engineering (now the Polytechnical School [Escuela Universitaria Politécnica-Eskola Politeknikoa], having added new degree programs) in 2002 when it obtained EUSKALIT's Silver Q, meaning that it earned more than four hundred points on an assessment of compliance with the EFQM model by a committee of external assessors.

As a consequence of the work done by all the members of the network in that academic year, a document was produced that was circulated to all the other centers as a guide to step-by-step progress on this path toward management improvement at a university center.

In the RCGI's third year, the 2003/2004 academic year, four new centers were added, for a total of ten centers. The new centers were the Leioa Faculty of Science and Technology (Facultad de Ciencia y Tecnología-Zientzia eta Teknologia Fakultatea) for the Bizkaia campus; and the Eibar School of Technical Industrial Engineering (E.U. Ingeniería Técnica Industrial-Industria Ingeniaritza Teknikoko U.E.), the Donostia–San Sebastián Business School (E.U. de Empresariales-Enpresa Ikasketen U.E.), and the Donostia–San Sebastián Faculty of Psychology (Facultad de Psicología-Psikologia Fakultatea), for the Gipuzkoa campus. During this academic year, a "Commitment to Excellence Diploma" was signed by four network centers, committing them to seek EFQM recognition in the near future (two or three years): the Faculty of Science and Technology (Bizkaia), the Higher Technical School of Engineering (Bizkaia), the School of Engineering (Araba), and the Nursing School (Araba).

Today, they continue to move forward with a structure and results manifesting this commitment and have the clear aim of evolving toward models of excellence in management; concretely, the EFQM model (figure 7.2) has been adopted as a reference.

Figure 7.2. EFQM Excellence Model

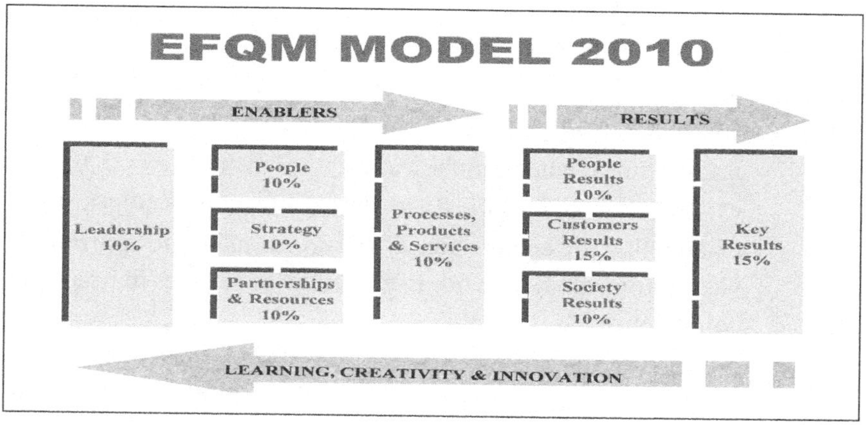

The 2004/2005 edition of the RCGI was called a "special edition" because it was specifically tasked by the office of the vice rector with defining the network's reason for being (who, what, why, and for what purpose).

- *Mission*

  RCGI is a group of UPV/EHU centers that seek to engage in quality management, taking the EFQM Excellence Model as a point of reference, supporting one another, sharing best practices, and working together to improve their management every day, encouraging the entire university organization to participate and cooperate in making the management of our entire university system a comprehensive shared project."

- *Vision*

  "The RCGI seeks to contribute to attaining the quality-management strategic objectives of each member center, while at the same time generating a reference model for management and organizational learning for all our university centers, with the goal that the UPV/EHU be managed using quality parameters recognized as standard in the European Union."

- *Values*
  - Continuous Improvement
  - Cooperation
  - Teamwork
  - Accountability
  - Commitment
- *General Lines of Action*
  - Supporting the implementation of the EFQM Excellence Model and of the quality plans
  - Facilitating organizational learning based on the exchange of experiences and best practices among university centers
  - Channeling the centers' needs and concerns to the appropriate vice rector's office, providing guidance for policies in its area of competence
  - Publicizing the achievements attained by the network centers to all the university centers, to the rector's administrative team, to the university community, and to society in general

The network's 2005/2006 edition culminated in the achievement of EUSKALIT's Silver Q (more than four hundred points on an external assessment of compliance with the EFQM model) by the Leioa Faculty of Science and Technology (Bizkaia) and the Vitoria-Gasteiz School of Engineering, marking a new stage in the progress of innovative management and the commitment to the EFQM Excellence Model by the RCGI member centers.

The next year, 2006/2007, saw a new benchmark with the admission of three new centers from the Bizkaia campus and one from Gipuzkoa, for a total of fourteen centers.

In 2007/2008, two new centers were added, for a total of sixteen (more than 50 percent).

The Program to Foster Teaching Innovation (Programa para el Impulso de la Innovación de la Docencia/Irakaskuntzaren Berrikuntzarako Programa, IBP) was developed to provide quality assurance in designing the new degree programs and adapting them to the requirements of the European Higher Education Area (EHEA).

At the same time, ANECA Internal Quality-Assurance Systems were designed and submitted to external assessment by two RCGI centers:

the Barakaldo School of Technical Mining and Public Works Engineering (E.U. de Ingeniería de Minas y de Obras Públicas-Meatzeen eta Herri Lanen Ingeniaritza Teknikoko U.E.) and the Donostia–San Sebastián Nursing School.

The Donostia–San Sebastián Faculty of Psychology and the Donostia–San Sebastián Nursing School obtained an EFQM-EUSKALIT Commitment to Excellence Diploma, and the Bilbao Higher Technical School of Engineering and the Vitoria-Gasteiz Nursing School underwent EFQM external assessment.

Throughout this stage, there were also a number of initiatives directed toward obtaining ISO-9000 certification, enabling us to learn how to implement a quality-management system with this in mind. The first certifications were obtained under the earlier ISO-9001/1994 standard, and subsequent ones were obtained under the new ISO 9001/2000 standard.

Experience was accumulated both in the area of the management of university services and in the management of teaching itself, with a frankly ambitious scope in some cases, as in the case of the School of Nautical Studies.

Some Examples

These are some examples of the activities to which we are referring:

- The Donostia–San Sebastián Nursing School, through a 2001 Asociación Española de Normalización y Certificación (AENOR [Spanish Association for Standardization and Certification]) audit, obtained ISO 9001/1994 certification for its training process management system for the diploma degree program in nursing.
- The Higher Technical School of Nautical Studies, through an AENOR audit dated October 15, 2003, obtained ISO 9001/2000 certification for its training process in all its degree programs.
- The Higher Technical School of Engineering, through an AENOR audit dated October 6, 2003, obtained ISO 9001/2000 certification for its management of business practices and organization of seminars, job fairs, and firm presentations.
- The School of Technical Mining and Public Works Engineering will soon obtain ISO 9001/2000 certification for its management of business practices.

The new regulatory framework arising out of the European Higher Education Area has established that universities should have formally established and publicly available Internal Quality-Assurance Systems (IQAS) and policies. For this reason, ANECA, together with the Catalan Agéncia per a la Qualitat del Sistema Universitari de Catalunya (AQU [Catalan University Quality Assurance Agency]) and the Galician Axencia para a Calidade do Sistema Universitario de Galicia (ACSUG [Agency for Quality Assurance in the Galician University System]), have developed the AUDIT Program, with the idea of providing guidance to universities in the design of their own quality-assurance systems.

The objective is to implement a cycle of continuous improvement in all processes and all spheres of university activity, making it possible to analyze the state of each process at any point in time, detect its aspects subject to improvement, implement actions that may enable progress, follow up on those actions, and determine whether they have had the expected effect, analyzing the new situation and continuously improving it.

Quality assurance should be understood as systematic, structured, and continuous attention to quality for the purpose of maintaining and improving it. In the context of this program, a proposed quality-assurance system should enable the center to send to both internal and external actors a clear signal of its commitment to actions ensuring and demonstrating quality.

Two UPV/EHU centers have participated in this new program: the Donostia–San Sebastián Nursing School and the Barakaldo School of Technical Mining and Public Works Engineering. The Internal Quality-Assurance Systems designed by these centers have been positively assessed by the UNIQUAL Quality Evaluation and Accreditation Agency of the Basque University System (Agencia Vasca para la Calidad de las Universidades del País Vasco-Euskal Unibertsitate Sistemaren Kalitatea Ebaluatu eta Egiaztatzeko Agentzia, UNIQUAL).

The Needed Changes in Organizational Culture

According to Francisco López (1997), universities have a hierarchical and bureaucratic (rule-based) culture and one that is internally oriented, or focused, on infrastructure improvement, assessments, processes of control, and systematic solutions to problems.

As a consequence of a study carried out by a large and multidisciplinary research team to which the authors of this chapter belong, we can affirm that our university, at least, has an organizational culture that is

Figure 7.3. Continuous Improvement in ANECA University Education

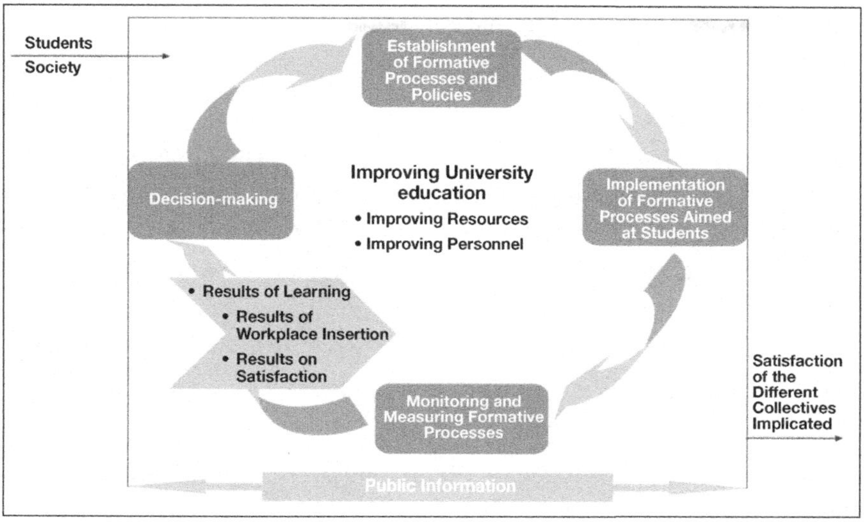

chiefly of the "clan" type and that is in any event internally oriented; that is, it is an "egocentric" organization.

The study focused on the sphere of research activities and found some indications of slippage toward organization of a "hierarchical" kind.

Based on our own observations and experiences, that part of the university organization whose area of activity is administration and services fits the "hierarchical" model much better. It is composed of individuals (administrative and service staff) whose terms of employment are very different from those in the academic sphere (teaching and research staff), and its organizational structure is clearly hierarchical and always oriented internally to the organization.

In sum, our university organization corresponds to an internally oriented, "egocentric" organizational culture, which in some areas shows "hierarchical" aspects and in others tends towards "clan" behavior.

At the same time, times change, and everything around us indicates the need for a "culture change." The twenty-first-century university is a social institution that has to concern itself not only with knowledge and erudition, but also with society's present and future needs. This requires an external orientation, and this is a radical shift, if we accept what has been said above. Many will continue to resist accepting even that the word "market" might have a role to play in university management, but if we simply replace it by "society," we can probably identify ourselves better and open ourselves to the external environment.

Conclusions

The creation of a professorial chair in quality is important as a driver of the "process of change," a point of reference both for developing and diffusing the principles of quality management and teamwork and for coordinating the quality programs promoted by the quality-management offices of the three UPV/EHU campuses.

The choice of the EFQM Excellence Model, promoted by the Basque Foundation for Quality-EUSKALIT and the Basque government, as a reference model for management.

Training in tools and especially in values for "teamwork," put into practice in numerous experiences promoting improvement teams.

The configuration of the Network of Centers of Innovative Management (RCGI), where best practices are shared, experiences are exchanged, and innovation, the learning process, and continuous improvement are promoted, taking the EFQM Excellence Model as the point of reference.

Sharing best practices in center management, by sharing good and not-so-good experiences, as well as the difficulties encountered in implementing the EFQM and ISO reference models and other actions promoted by ANECA, always with the idea and the attitude of "learning from others" and "continuously improving."

The progress made in the necessary change in "organizational culture" from a clan model and one that is very much internally oriented to one oriented toward the market and society—that is, one that is ever more clearly open to the external environment.

References

Ayestarán, Sabino, and Aitor Aritzeta. 2003. Aplicabilidad de la teoría de los roles de equipo de Belbin: Un estudio longitudinal comparativo con equipos de trabajo. *Revista de General y Aplicada* 56 (1): 61–75.

Cameron, Kim M., and Robert E. Quinn. 1999. *Diagnosing and changing organizational culture based on the competing values framework.* Reading, MA: Addison-Wesley.

European Foundation for Quality Management (EFQM). 2010. *Modelo EFQM de excelencia.* Madrid: Club de Excelencia en la Gestión.

López, Francisco. 1997. *La gestión de calidad en la educación.* Madrid: La Muralla.

Panera, Francisco. 1999 (January). Calidad total en la enseñanza universitaria. *Revista Dirección y Organización de la Fundación Universidad Politécnica de Madrid* 21: 94–102.

Panera, Francisco, and Sabino Ayestarán. 2008 (January–February). "La gestión de calidad en la Universidad del País Vasco—Euskal Herriko Unibertsitatea." *Revista Forum de Calidad de Madrid* 20: 37–44.

Scholtes, Peter R. 1988. *The team handbook: How to use teams to improve quality.* Madison, WI: Joiner Associates Incorporated.

Senge, Peter. 1994. *The fifth discipline fieldbook: Strategies and tools for building a learning organization.* New York: Currency, Doubleday.

West, Michael A. 2004. *Effective teamwork: Practical lessons from organizational research.* 2nd ed. Oxford: Blackwell.

8

The Technological "Interface" in Contemporary Art

Josu Rekalde
Translated by Jennifer Martin

It is obvious that the technological change we live with today is changing how we behave on a deeper level than what Marshall McLuhan himself could have imagined in the 1970s. At the beginning of the twentieth century, the term "technology" evoked images from the nineteenth century: heavy cast iron axles and gears that rotated and squeaked at the mercy of a thermal impulse from coal or oil; the car, the motorbike, and the airplane would later progressively replace these symbols of technological progress, arriving at the electrical and electronic age where the television prevailed as an icon at the end of the twentieth century. In contrast, when people ask us about technology these days, the only thing that comes to mind is a small object full of microprocessors that silently and invisibly manages infinite automated processes, displaying everything on a beautiful color screen and in high-resolution graphics, of course.

The use of hands has, for human thought, signified its first intermediary with external objects, its first interface with reality; this communication task established between thought and the ability to transform objects was the seed of the technological process; and this process, along with the appearance of language, would transform the mental structure itself of human beings. With information technology, we have gone from using the hand as a direct instrument to "manipulating" the physical environment with remote-controlled electronic interfaces and teleprocessing. Despite the fact that in the measure of a human life, a relatively large amount of time has occurred between these two stages, it represents a relatively small amount of time with respect to the evolution of the species.

With the entrance of electronics in the art world, interface began to act as an intermediary between the image and its observation. As Peter Weibel (1999) says: "The current horizon of the creation and contemplation of images has been replaced by an artificial horizon permanently dominated by interface technologies."

We have to ask ourselves about the needs and motivations that have led to this transformation, about the implications of social change in technological change. The technologies did not arise from the abstract study of a team of inventors, but were born in order to satisfy social needs, and arose instead as answers to questions that a group of people asked: "In human ecosystems, technological change is a frequent response to diminishing returns" (Harris 1980, 192).

In the industrial age, art seemed to be limited to a few risky pieces that were more or less successful. The art market survived by keeping the same Renaissance technology, changing only the form of representation. Technologies like photography influenced those representations. On one hand, freeing painting from having to fulfill an analogous role with reality, and on the other hand, inspiring research into phenomenon like the retinal perception of color. But the true avant-garde revolution did not lie in the forms created and expressed in a painting or a sculpture, but instead in the ideological fact that it enabled a great number of authors/creators to have access to artistic production. They broke away from the unionism of the artistic trade to a certain extent, in this way depleting the previous production systems and demanding in art, as happened in other production systems, the consequent technological change.

Technological Interface in Art

The past century began with speed as a reference to modernity; technological progress became the aesthetic symbol. Thus went the fourth postulate of the proclamation that Filippo Tommaso Marinetti made in the Futurist Manifesto (1909; 2006, 13): "We believe that this wonderful world has been further enriched by a new beauty, the beauty of speed. A racing car, its bonnet decked out with exhaust pipes like serpents with galvanic breath . . . a roaring motorcar, which seems to race on like machine-gun fire, is more beautiful than the Winged Victory of Samothrace." This expression, born as a reaction to the post-romanticism of the late nineteenth century, was characterized by the worship of modernity, the machine that advanced at full speed, including the war machine that was being prepared for World War II.

Within this context, the Italian futurists and the Dadaist painters became quickly interested in cinematography, a technology that had already been in use for twenty years. The influence of chronophotography by Eadweard Muybridge and E. J. Marey was evident in certain paintings such as *Nude Descending a Staircase* by Marcel Duchamp (1912). Some studies revealed this cinematographic influence, even in poetry, where the words "in freedom" became an art in motion, juxtaposing their meanings, one to the other, in the same way a montage does.

The television screen or monitor appeared in the 1950s as a mosaic of light dots that responded to electronic signals arranged across Cartesian coordinates and to a representation system as old as that of the Renaissance perspective: in short, a two-dimensional plane in which we could depict and create with our ideas. The electronic matter that we worked with could house channels of dazzling images in its mosaic, as well as spatially and temporally transform all that fell within this network. This ductile material, from the spatio-temporal point of view, implied a new experimental pole for art in which images no longer remained fixed in a spatial framework, but were in constant metamorphosis.

Space and time, inseparable concepts in principle, accorded greater room to maneuver in this case. Situating ourselves toward the spatial pole, we could work in 3-D, which meant a video installation or video sculpture. Situating ourselves toward the temporal pole, we could have all those works that emphasize concepts like linear rhythm or cyclical rhythm, direct and delayed time, *feedback*, and so on. Chronometric time as a social measure and chronometric time of the electronic piece itself (twenty-five images/s) were nothing more than mere approximations to the temporality that the author expressed and that the observer perceived.

In the 1970s, a FLUXUS artist group that included John Cage, Joseph Beuis, Nam J. Paik, and Wolf Vostell, among others, experimented with the videographic medium. In 1963, Nam J. Paik presented an exposition called Experimental Television, in which he disrupted the flow of electrons in a cathode-ray tube from a television, receiving the first abstract television sequence live, without needing the intermediation of the camera or the video tape recorder. In this exposition Paik declared: "My experimental TV is the first ART (?), in which the 'perfect crime' is possible . . . I had put just a diode into opposite direction, and got a 'waving' negative television . . . My TV is NOT the expression of my personality . . . My TV is more (?) than the art, or less (?) than the art" (Paik 1963; 1993).

In addition, Wolf Vostell had used television monitors in installations since 1959, but unlike Paik, they were symbols in and of themselves. Regardless of the image that it conjures up, the television was, along with the car, the most powerful symbol of Western culture at the end of the twentieth century. Vostell believed that media art left a mark on our lives in the same way as a drawing. His work consisted of the application of a *decollage*, which contrary to the *collage* did not involve pasting one image over another but instead tearing off the top layers, tearing off part of one image in order to see parts of the images underneath, like a metaphor of the culture in which many superimposed sociopolitical layers coexisted and that the artist was responsible for revealing.

At the same time in 1968, Jim Newman and Gerry Schum founded the first galleries specializing in video art. The former organized the Dilexi Foundation, while the latter established the Video Gallery Schum, with works from Land Art, such as those from Richard Long and Walter De Maria. From that moment on, video that had been born with an activist component against television, turned toward artistic circuits that were converting themselves into another artistic movement: video art.

A second generation of artists emerged, experimenting with the medium of technology as an instrument for posing questions about human relationships and language itself. This was the case of Bill Viola and Gary Hill. Bill Viola, without a doubt the most influential video artist, has used the video camera in an autoscopic way from the beginning, studying the behavior of the physical body in the most unusual situations. He locked himself in an empty house and remained there without sleeping, and in total isolation for three days and three nights, to create his work *Reasons for Knocking at an Empty House* (1983). Subjected to the watchful eye of the camera like a prisoner in his cell, his breathing, any movement, even swallowing, increased in significance as acts in which self-knowledge and self-awareness were tested, along with the capacity for suffering.

In the 1990s, a group of women artists emerged in the international panorama with the common characteristic of creating intimate performances with the sole presence of the camera. This group's works were based on the use of video as a tool for a reflection on gender identity. In this way, artists such as Sadie Benning, Pipilotti Rist, Michael Curran, and Cheryl Donegan displayed a series of events onscreen that left the spectator trapped between a passive look and the game displayed for viewing on the screen.

Even though it started a bit later, videographic work done in the Spanish state followed the same steps as in the international context. The first videographic events in Spain were immersed in an effervescence of artistic activism, in which video was just another instrument, although a very powerful one due to its immediacy and to its dissemination capacity. What was important was the act itself: The artistic concept was there, and video could record and duplicate it as many times as necessary, without the intermediation of institutional filters. The happenings and performances were converted into political acts daily on the streets; some artists, such as Antoni Muntadas, Francesc Torres, Joan Rabascall, and Eugenia Balcells, knew how to reflect, from a medium such as video, upon the media aspect of their own culture. This artistic activism involved an eagerness for social transformation that still coexisted with that of other countries but also acquired very different leanings at the end of the Franco era.

The Iruñea-Pamplona Encounters (Los Encuentros de Pamplona) of 1972 were a meeting point and a catalyst for innovation in contemporary art in the Basque Country. These meetings, which were largely due to the impetus of music composer Luís de Pablo and to the patronage of Juan Huarte, opened up the door to arts in action and to electronic arts. Special mention was made of the musician and artist John Cage, a member of FLUXUS, in these meetings. With regard to Basque artists, it should be highlighted above all that political conflict caused the Spanish government to censor some of their works (during the final years of Franco's dictatorship). Luis de Pablo's electro-acoustic music and the performances of Esther Ferrer and her group Zajwere were decisive in terms of innovation.

Video festivals began in the 1980s. In 1982, a video festival that would be groundbreaking in Spain was born: the Donostia–San Sebastián Video Festival (Festival de Vídeo de San Sebastían). This festival developed as an appendage to the city's film festival. There were only three editions of the festival, owing above all to the shock of coexisting under the weight of the media surrounding famous cinematographers. Later, the Vitoria-Gasteiz festival was born in 1984, and the Tolosa festival was first held in 1986.

The first wave of artists that used video in the Basque Country came together precisely at the Donostia–San Sebastián Video Festival during the three sessions that the festival held, from 1982 to 1984. The festival was run by Guadalupe Etxebarria and had three subsections: the competitive section, an international panorama of video screenings, and a section for conferences and live action. Synchronously with the Donostia–San Sebastián Video Festival, the first students specializing in audiovisual media graduated from the School of Fine Arts in the Basque Country (whose first students graduated during the 1982–1983 academic year). Multidisciplinary artists such as Juan Carlos Eguillor (with his video "Bilbao la muerte" or "Bilbao death"), the director José Mari Zabala, Gerardo Armesto, Armando Dolader, and Antonio Herranz presented their work in these first festivals. Together with these artists, a generation of new authors focused on videographic creation, airing videos and installations from Iñigo Salaberria, Isabel Herguera, Mikel Arce, Josu Rekalde, and Francisco Ruiz de Infante.

In the 1990s, even though video was already a reality in most expositions, the artists' curricular background worked against them when it came time to view and distribute the work. The video art circuit had practically disappeared. All those authors who came from that scene and wanted to remain in the artistic field had to retrain themselves and retreat into the trends of the art market, which in the Spanish state, had not known how to draw collectors to the video artists' sector. This evolution, partly induced by the fact that video artists kept their own distribution circuits away from the art object circuit, ended in the Internet era, in limitless commercial diffusion.

A new wave of artists became known in video from a more casual perspective and was less attached to the medium. While the videos of the 1980s were characterized by poetic introspection, in the 1990s, authors such as Jon Mikel Euba, Sergio Prego, Itziar Okariz, and Javier Pérez approached video without any intention of dominating it on a technological level or by means of language, only using it to document an artistic project where the atmosphere in which the author was the protagonist mattered most. In this respect, video went from being an internal self-portrait to converting itself into a self-portrait of the body itself and of its identity in the 1990s. This self-portrait or self-copying characteristic was perhaps the best definition of what video as a tool had given and was giving to the artists.

Multimedia, Navigation, Interactivity, and Art on the Web

In art, the term "multimedia" essentially referred to the interdisciplinary work in which visual, audible, tactile, and temporal forms were combined; thus, painting and sculpture fused together with literature, video, film, and music. Later, this term also became applicable to the computer, showing the connectivity and interconnection of different visual and audio media. In the 1990s, the CD-Rom was the most popular multimedia format, but from today's Internet perspective, it is a consolidated reality. The CD-Rom was a testing ground, where interactivity, multimedia navigation, and hypertext were developed.

Interactivity is one of the key concepts in art that is currently seen with new technologies; this concept is trying to abandon the old idea of unidirectional communication established by Jakobson's outline, where the Transmitter delivers a Message through a channel and the Recipient passively receives it. With interactivity, a reciprocal path is established between the receiver or spectator and the message sent by the transmitter or artist. Using communication theory terms: a new channel that picks up feedback is established. This type of work offers the spectator the possibility of not just following one path of observation, but of actively choosing and influencing a whole series of paths that the author has predesigned as a network of elements.

Depending on the strategies marked by the author, we are talking about *Navigation* when there is interactivity in which we can advance or use the computer's mouse to choose the path we follow from a page or image that is presented as a menu; or we are talking about *Immersion* when the choices are intuitively made with an interface that is based on our own senses as a prosthesis. There is no doubt that today the Internet is a great virtual space to which we are all connected and where many interactive experiences become visible. In this sense, art has been on the web since the beginning. The terms *Net Art* and *Art on the Net* or *Internet Art* were quickly invented. Basically, pieces of art specifically created for the Internet were called Net Art, which in its early days turned out to be very important as it differed from art that used the Internet only to exhibit its own activity.

In the early days of the Internet, artists and thinkers that were drawn to it dreamed of a "space of freedom" without any type of control, a place where the ideas and artwork could flow without filters. Later, it began to be thought, as in the case of the artist Mariano Maturana during the video meetings of Iruñea-Pamplona in 1996, that the Internet represented a temporarily free space to establish thought networks, making it necessary to take advantage of the indeterminate control (that currently exists) to create these human networks.

Transformation Technology in Real Time, Virtual Reality, and Augmented Reality

During the past few years, due largely to the video game industry, software and hardware have been designed to create spatial and audio transformations in real time.

Virtual reality has a technological interface that acts by immersion and in some way replaces our senses, giving the impression of actually being inside of digital and virtual space. These interfaces range from data gloves or stereoscopic glasses (3-D glasses) to motion simulators and exoskeletons that completely surround the visitor's body.

Analyzing it from the concept, these technologies that are very embedded in the video game industry feed on very simple ideas, such as not depending on the limitations of the physical world where gravity exists. In this sense, virtual reality seeks the release of the human body from its bonds, like an updated utopia that finds its greatest ally in technology. The body cannot fly nor pass through solid spaces, but with the help of a computer image, its iconic ghost does just that, interactively synchronized to its movements. This splitting between the physical body, fooled by a framework of simulating equipment, and its ghost image makes us think of the cultural schizophrenia in which our civilization is submerged, because the research in this field tends to entirely confuse the body's actual position and its physical reality with the programmed image of a body free of ties. The release resembles the splitting of the body and the soul, which according to some religions, happens when the physical body expires.

A technology that is more attached to the physical world is called *augmented reality*. Unlike virtual reality, augmented reality does not replace physical reality, but adds computer data to the real world, combining physical information with virtual information devices. The maps managed by GPS devices are a classic example of augmented reality. Information about the location that we actually view is added to them. There are many examples in the art field where this type of technology is at the center of projects. Some of them are like those of the Situationists, devising psychosocial paths through cities or strategically expanding the view of a location displaying conflicts.

Contrary to what they want to sell us with interactive games and representations of current electronic iconography, where with the help of machines we can fly and are virtually liberated, in the field of immersion, the artists' suggestions are aimed at spectators being conscious of their real dimension without deceptions, but being able to use the technology as an extension of their senses.

During these last few decades we have witnessed the digital convergence of media that came from telecommunications and art's own media such as recording, photography, and film. All this digital convergence brought with it a new way of understanding creation; we went from understanding visual creation as recorded or archived, to the possibility of visually creating in real time, and thus bringing us closer to musical creation environments. Just as the filmmaker Peter Greenaway showed in Bilbao in 2008, the future of the audiovisual as language is on cellular phone screens and on the Internet, it is through the portals of *Second Life*, where at the moment, they are producing stories that young people identify with.

New Technologies, Old Myths

The political fiction novel *Nineteen Eighty-Four*, written in 1948 by George Orwell, conveyed a negative utopia over the social use of control through new technology. Our homes today, places of work and leisure, are invaded by security cams, web cams, life cams, and virus spies *(spyware)*. All these technologies, connected in real time to telematics networks, form a control and a serious danger of being unprotected at the same time. But from the ideological point of view, the idea of control through an interface does not differ much from the Renaissance concept of visual perspective where the viewer's gaze remained in the center, dominating the universe. We live in a monitored world in which all data can be displayed: health, the economy, politics, even the relationships of couples are mediatized through the screens.

Our gaze has become nonconformist and wants to view more: more through film, more through television, more through the Internet; the "voyeuristic" view is the greatest exponent of our current unfulfilled desire. Art generated the symbolic world before; now the mass media and advertising generate it. The museums in this context, heirs to an elite bourgeois taste, seem to have no other option than to approach the media spectacle that surrounds theme parks. Behind the theatrical stage machinery, art is directly searching within the spectacle for a dignified exit. In the twenty-first century, just like in the beginning of the twentieth century, fusions occur more and more frequently between visual artists, recording artists, theatrical artists, and choreographers in order to investigate an imaginary environment closer to the spatio-temporal one than to the one in which we belong. The video jockey (VJ) culture is an example of this.

The institutional assimilation of artistic productions accomplished through media technology always involves a delay. A fitting example of this is that the first time a videographic work was accepted in the awards convocation of the artistic creation Gure Artea, the biannual art awards supported by the Basque government, was in 1996.

With art mediatized by interface, new ways opened up to think about the old existential questions. In an age when science began to put spatial and temporal limits on the universe and its expansion, the physical body with all its limitations seems to blur in an upheaval where its identity is mistaken with its representation. Many artists currently work with the new dimension that technology can contribute to the concept of identity, stretching the limits imposed by nature. This idea, far from being a reality, has become a symbolic image shared by culture and the Cyborg fad. It would be interesting to open up an ideological debate in the heart of the artistic community, without historical complexes and without technocentric pretexts. In art, to think of a place to experiment is to think of a place to grow personally; it is rethinking creation and the appropriate interpretation and pedagogy of art, like a laboratory for cognition. The media and connectivity provide the largest present-day iconographic source; they are the new cathedrals and the new museums. But the body of the author, of the creator, of the spectator, of the student, and of the professor are physically equal to body of the artist that painted the bison in the prehistoric caves of Altamira, Lascaux, Santimamiñe, or Ekain. Like them, we turn to thought and to symbolic intermediation when designing and sublimating our joys and fears.

References

Baigorri, Laura, and Lourdes Cilleruelo. 2006. *Net Art: Practicas artísticas y estéticas de la red*. Barcelona: Brumaria.

Bonet, Eugeni. 1993. *Desmontaje*. Donostia: Arteleku.

Bonet, Eugeni, Joaquim Dolls, Antoni Mercader, and Antoni Muntadas. 1980. *En torno al vídeo*. Barcelona: Gustavo Gili.

Cilleruelo, Lourdes. 2001. *Arte de Internet: Génesis y definición de un nuevo soporte artístico tecnológico*. Bilbao: UPV/EHU.

Deleuze, Gilles. 2005. *The Movement-Image*. Translated by Hugh Tomlinson and Barbara Habberjam. London and New York: Continuum.

Gonzalez, Ander. 1995. *El vídeo en el País Vasco (1972–1992)*. Bilbao: UPV/EHU.

Hamaca. www.hamacaonline.net.

Harris, Marvin. 1980. *Culture, People, Nature: An Introduction to General Anthropology*. 3rd ed. New York: Harper & Row Publishers.

Marrinetti, Filippo Tommaso. 2006 [1909]. "The Foundation and Manifesto of Futurism." In *F. T. Marinetti: Critical Writings*, edited by Günter Berghaus and translated by Doug Thompson. New York: Farrar, Strauss, and Giroux.

Medien Kunst Netz. www.medienkunstnetz.de/source-text/31/.

Mitry, Jean. 1997. *The Aesthetics and Psychology of the Cinema*. Translated by Christopher King. Bloomington: Indiana University Press.

Moles, Abraham. 1974. *La comunicación y los mass media*. Translated by Juan José Ferrero. Bilbao: Editorial Mensajero. French: 1973. *La communication et les mass media*. Paris: Les Dictionnaires Marabout Université.

Paik, Nam J. 1963; 1993. "Afterlude to the EXPOSITION of EXPERIMENTAL TELEVISION." In *In the Spirit of Fluxus*, organized by Elizabeth Armstrong and Joan Rothfuss. Minneapolis: Walker Arts Center.

Popper, Frank. 1993. *Art of the electronic age*. London: Thames and Hudson.

Rekalde, Josu. 1988. *Bideo artea Euskal Herrian*. Donostia: Kriselu.

———. 1995. *Vídeo, un soporte temporal para el arte*. Bilbao: UPV/EHU.

Rekalde, Josu, Lourdes Cilleruelo, Agustín Ramos, Arantza Lauzirika, Josu Aguinaga, and Iñaki Billerabeitia. 1997. *Lo tecnológico en el arte, de la cultura video a la cultura ciborg*. Barcelona: Editorial Virus.

Various Authors. 1996. *Primeros Encuentros de Video de Pamplona. Actas de los primeros encuentros de video de Pamplona editado por la Institución Príncipe de Viana*. Pamplona: Príncipe de Viana.

Vicario, Begoña, and Jesús María Mateos. 2007. *Sistiaga, el trazo vibrante.* Animadrid Collection No. 3. Madrid: Consejería de Cultura y Turismo de la Comunidad de Madrid.

Villota Toyos, Gabriel. 1977. *Videoacción: El cuerpo y sus fronteras.* Exhibition catalogue for "Luces, cámara, acción (…) ¡corten!" Valencia: Institut Valencià d'Art Modern (IVAM).

Weibel, Peter. 1999. *Conocimiento y visión.* En el Catálogo Cibervisión 99. Muestra Internacional de Arte, Ciencia y Nuevas tecnologías, Madrid, November 16–21.

9

Innovation and Creative Content in Internet Media

Ainara Larrondo Ureta

Translated by Jennifer R. Ottman

In recent years, creativity and innovation have become two indispensable concepts. Even if both are associated with originality, imagination, and talented productivity, creativity alludes to the ability to conceive novel ideas and create something that did not exist previously—at least in its totality—while innovation refers to the application of those ideas (Nadal and Román 2008).

The World Wide Web has represented and represents one of the chief contexts for creativity and innovation in the communicative sphere. Its discovery in the mid-1990s entailed a full-scale upheaval for the media, especially for the press, which was traditionally slow to change. This sector considered the web a medium with significant impact and a great capacity for attracting younger audiences, and hence a novel tool capable of offering business opportunities. In the end, launching the first online editions did not cost much money, did not endanger the sales of printed newspapers—the number of Internet users remained laughably small—and was a fashionable hallmark of modernity about which publications could boast, in the face of the exhaustion of traditional formulas. In this regard, the web was not initially distinct from other earlier platforms, where attempts had also been made to respond to the demands for innovation and technological evolution arising from the very fact of the media's survival.

All the same, in this first stage, known as "shovelware" (Thalhimer 1994, 46–47), the content of online editions showed a real lack of creativity, being based on already published materials that were recycled and adapted for virtual distribution. Online products were clearly homogeneous, and it was still difficult to speak of a new language, although media websites began to demonstrate (timidly) certain advantages, such as constant updating, personalization of content, and nonlinear narration, building on communicative elements foreign to traditional journalism, such as hypertext links.

This panorama changed qualitatively and quantitatively in only a few years, at the end of the 1990s, as a consequence of the exponential growth in the number of online media—with and without offline counterparts—and of the greater interest among journalistic enterprises in taking advantage of the Internet's potential. Among other decisive initiatives, the first online editorial teams were set up, mostly made up of young people eager to experiment, enabling a stage of maximum exploration and development of new journalistic formats. In this evolving context, some authors viewed the first generation as having already come to an end (Deuze 2001).

Although fears for the viability of journalistic enterprises on the Internet remained, a period of transition and adaptation to the expressive possibilities of the online medium thus opened at the dawn of the new century. The special coverage of the September 11 terrorist attacks in New York was a turning point in this regard because it led the large media companies to demonstrate their creative spirit and put effort into developing truly innovative content, produced with full awareness of the journalistic uses of the Internet and of its particular tools. The young online media then found themselves forced for the first time to put their better technological response, greater editorial consolidation, and better use of multimedia to work in covering major events and promoting access to information through the Internet (Salaverría 2005, 85–86).

In these years also, voices discussing the convergence initiatives promoted by some media became ever louder. As Martha Stone (2002) of the Innovation International Media Consulting Group recognized at the time, single-medium companies on every continent were transforming themselves into multimedia companies, integrating the editorial operations of their audiovisual, print, and web divisions. Concretely, 73 percent of the members of the World Association of Newspapers (WAN) acknowledged at the time that convergence of some kind was emerging in their companies.

Throughout the century's first decade, communications groups have increasingly felt the effects of multimedia convergence and the need to adapt to the changes imposed by this trend toward the integration of media, languages, and content. As they have increasingly become "content producers" capable of reaching their audiences with a ubiquitous, transversal, and multiplatform trademark, these groups are staking their efforts today on a strong sector that is competitive in multimedia formats, with effects on the routines, principles, and models traditionally used to distribute information.

New Content Models

The content sphere represents a real terrain of applied creativity (innovation) in online journalism. Online information products are based on novel ideas and represent something new, and this "something" is interesting, suggestive, and at a high level of quality for its audience. These efforts to maximize the profitability of novel formats mixing technology, codes, and the discovery of new forms of expression in order to give new value to web content are bearing fruit. The consumption of journalistic media has become one of the chief activities of Internet users, and the demand for online information is increasing exponentially. The numerous reports and statistics available on the subject indicate that the Internet will be the top source of information in the next decade, a forecast that might explain the continuous growth of investment in online advertising and the amounts staked on this new model of journalism.

On the other hand, these new contributions have been recognized as such and incorporated into journalism's system of knowledge, in response to the problem of adapting content to the new medium. What's more, this assimilation has generated major changes and a rupture in the previously known system of genres, as is demonstrated by the classification of this content and its forms developed by the academy. This has also forced a rupture with regard to research procedures. In this regard, the issue of methodology has become in itself another of the most prominent challenges of online journalism, a terrain equally friendly to innovation on the basis of new approaches and techniques of content analysis.

Linguistic innovation for the purpose of discovering new forms of storytelling has represented one of the chief challenges faced by Internet journalism from the beginning. A decade ago, Jonathan Dube (2000) published a complete repertory of some of the most novel modes of storytelling that the chief media websites were beginning to put into practice on the Internet. Today, there exists a variety of examples of innovation in online informational coverage, narrative models that embody maximum creativity in this environment: audio slideshows,[1] photo and video essays,[2] 360-degree panoramic videos,[3] interactives,[4] interactive maps,[5] and Flash or multimedia packages (infographics and special reports).

Today's young people are tomorrow's users and are used to audiovisual, playful codes, leading the media to attract their attention through formats that incorporate these codes into the informational discourse. Together with the philosophy of "Show, don't tell!," the philosophy of "Interact, don't show!" has begun to be forcefully applied, and spaces based on interactive[6] and immersive[7] environments (virtual-reality or simulation environments), similar to those used in computer games, have begun to spring up. One of the first products of this kind was the immersive "Can You Spot the Threats?"[8] published by *MSNBC.com* in 2001, in connection with the September 11 attacks. The game simulates an airport scanner and puts the user in the role of a police officer trying to detect possible security threats in passengers' luggage. These game-based models increase their audience's involvement by inviting them to undergo an experience that can enable them to think about the news and understand more exactly what the article is trying to convey.

---

1. www.guardian.co.uk/audioslideshows
2. *It's Not Easy*, SergioCaro.com (2008)
3. *Haiti Earthquake* (2010), edition.cnn.com/interactive/2010/01/world/haiti.360/index.html
4. *Primer cara a cara Zapatero-Rajoy* (2008), www.rtve.es/elecciones/debate-25-febrero.shtml
5. www.timesonline.co.uk/tol/news/uk/article3524866.ece. In 2008, *The Times Online* surveyed its readers about the state of the economy in the United Kingdom. Hundreds of people added comments about the economic issues that most concerned them, and the responses were included on an interactive map created for the occasion, in which Google Maps technology was used for the first time to illustrate and complete one of the newspaper's articles.
6. www.cnnchallenge.com; www.elmundo.es/especiales/2003/02/salud/genetica/descifrar_la_vida.html
7. www.immersivejournalism.com
8. www.msnbc.msn.com/id/34623505/ns/us_news-security

Those media that have been making use of modalities of this kind have seen their excellence recognized through specific awards, such as the Online Journalism Awards (Online News Association)[9] and the Eppy Awards for the Best Media-Affiliated Internet Services (*Editor & Publisher* and *Mediaweek*). These prizes reward investments in audience participation, originality of content, creative use of new multimedia languages, and effective navigation design. In sum, they provide a way to acknowledge the vitality and innovation of Internet journalism at a time of transformation in the industry.

Content Convergence: Multimedia

"Media convergence" has become one of the fundamental axes of current processes of innovation in communications, although it is not strictly a new phenomenon in itself. In fact, this fashionable term expresses the continuous transition—*convergence continuum* (Deuze 2004, 140; Dailey, Demo, and Spillman 2005, 150)—in which journalism has been immersed for decades, from nonconvergence to a state of multimedia convergence in which the hope is, sooner or later, to achieve integration of the parts of the productive process and editorial coordination across editorial teams working in different media. This gradual process is, in addition, multidimensional because it has simultaneous implications in the technological, business, professional, and content spheres[10] (López and Pereira 2010).

The convergence initiatives implemented recently have made this interrelationship evident, as well as the factors that have most conditioned content innovation up to now: new technologies and overcoming certain technical limiting factors (connectivity and bandwidth); planning at the editorial and management level; investment; the autonomy of editorial teams and their particular work cultures; and training professionals to experiment on the basis of their own creativity and ingenuity.

In this process of convergence, the media's arrival on the Internet has made it possible to take the definitive step from *single-medium* to *multimedia*, one of the chief changes to traditional communications paradigms experienced in recent years. While print uses static texts, photos, and

9. http://journalists.org/

10. A team made up of subgroups from twelve Spanish universities has taken on the study of these four aspects as part of the "Digital Convergence in the Communications Media" project (SEJ2006-14828-C06) (2006/2009) financed by the Spanish Education and Science Ministry (www.infotendencias.com).

graphics, television uses video, and radio uses audio, in the online environment the journalist can use any of these kinds of media to present his or her stories, or all of them at once, and has other possibilities, unknown on previous platforms, available as well:

> The Internet not only embraces all the capabilities of the older media (text, images, graphics, animation, audio, video, real-time delivery) but offers a broad spectrum of new capabilities, including interactivity, on-demand access, user control, and customization. Thus using the new media tools available via the Internet, online journalists can tell stories using whatever modalities and communication features are needed and appropriate for a particular story. (Pavlik 2001, 3–4)

In this advance toward content convergence, the advance from "*many media*" toward "*multimedia*" (Paul and Fiebich 2004) has also been decisive—from a language based on aggregating content in different media formats (hypermedia)[11] to one based on integrating this content for simultaneous consumption in a unitary message (multimedia). Hypermedia language entailed an entirely novel approach in comparison with the written language of print and the audiovisual language of radio and television, but a truly multimedia language is a much more exact representation of the divergent paradigm that online communication entails. Today, both models, hypermedia and multimedia, coexist and pose challenges to traditional journalistic writing, which depends more than ever on well-planned strategies of hypertextual and multimedia composition (Deuze 2001; Kawamoto 2003, 182).

Accordingly, media convergence in the sphere of online content has been reflected through experimentation with prototypes that take advantage, to varying degrees, of the potential of a convergent or multicode language, the maximum expression of which is found in multimedia products. These products may include already familiar textual, visual, audiovisual, and graphical codes (in addition to other novel elements such as 3-D, mash-ups, animation, and interactivity) applied to video, audio, and text. Behind a good part of this innovation in multimedia design, content,

---

11. This system makes it possible to relate autonomous fragments of information (nodes) in different media formats, through links and nonsequentially. Hypertext has come to be located at the point of departure for the discursive novelties of online storytelling in comparison with the traditional written story, and its use for content development brings with it more creative and less rigid formulas. This implies significant changes in classic patterns of newswriting and journalistic genres.

and genres is Flash, the Internet's most famous web design technology, created in the same year as the first online media enterprise.

This type of multimedia content may adopt the form of animated infographics, also known as "interactives." Since they began to develop on the Internet around 1998, these infographics have gone from being a mere translation of static graphics to forming a new genre, specific to the medium, capable of taking full advantage of the web's possibilities for textual, iconic, and sonic integration. For this reason, multimedia infographics demonstrate a kind of narrative that is distinct from any other, one that is three-dimensional[12] and supported by databases[13] and by maps in different formats.[14] As is evident in the interactive published by *Elpais.com*[15] on the 2007 elections in Spain's autonomous communities and municipalities, the content convergence promoted by Flash infographics can be the fruit of processes of cooperation between different media belonging to the same group—in this case, between the television network CNN+, the radio broadcaster Cadena SER, and the newspaper *El País*.

Many examples exist of good multimedia infographic productions on both sides of the Atlantic, including those that have been internationally recognized through the successive awards of the Malofiej International Infographics Prizes (Premios Internacionales Malofiej de Infografía), organized by the Spanish Society of News Design (Society of News Design España, SNDE).[16] For this reason, we refer the reader to the infographics chosen year after year as a sample of the cutting edge of this novel format, a symbol of the creativity and functionality inherent to online content.

Special Reports: An Example of Innovation

Informational mega-events are the best growth media for trying out new formulas for developing and presenting content, as the special reports produced by the leading online media outlets demonstrate. These reports

12. *Sacred Ground: The Building of the Pentagon Memorial* (2008), www.washingtonpost.com/wp-srv/nation/interactives/pentagonmemorial/victims/

13. *Election Results 2008: President Map* (2008), http://elections.nytimes.com/2008/results/president/map.html; *Home and Away: Iraq and Afghanistan War Casualties* (2010), http://edition.cnn.com/SPECIALS/war.casualties/index.html

14. *Haití, territorio devastado* (2010), www.elmundo.es/elmundo/2010/graficos/ene/s3/resumen_haiti.html. This graphic integrates illustrated maps, Google Earth views, and a temperature map.

15. www.elpais.com/especial/elecciones/grafico.html

16. www.snd-e.com/es/malofiej/premios

give new power to the informational and documentary functions that the "in-depth report" has fulfilled in its traditional written and audiovisual manifestations, making them one of the most innovative expressions of large-scale reporting on the Internet, even as they give birth to a native genre of online journalism.

This exclusive typology has up to now been an excellent testing ground for hypermedia and multimedia experiments, and hence one of the best examples of the kind of innovation that the Internet's convergent narrative has brought to journalism's classic genres. Equally, this model represents one of the best indicators of the potential of journalistic language on the Internet. Paradoxically, its analysis also demonstrates that innovation and creativity in the online medium have on occasion been subordinated to editors' choices in favor of less expensive and more flexible products, such as video.

In fact, special reports are complex; they demand time (days, weeks, and even months, depending on the subject) and technical and human resources (multifaceted professional teams made up of web designers, writers, documentary producers, infographics creators, camera operators, and on-air talent), all of which has contributed to the idea that they are expensive and not very profitable. What's more, products of this kind are the best evidence that innovation demands investment and a spirit of creativity, something that is not always possible in a context in which production tends to be standardized as much as possible, through reliance on reusable formats, standard web architectures, and the incorporation of tags in cataloguing reports. Despite everything, what is certain is that the benchmark media outlets continue to develop even more products of this kind.[17] Not coincidentally, they offer a good opportunity for claiming ownership of particular subjects and are synonymous with Internet journalism: They concentrate the advantages of online publication—hypertextuality, multimediality, and interactivity—in a highly dynamic format, thereby serving to give a more complete perspective on events and respond to the role adopted by news sites in the communications market, as suppliers of qualitatively and quantitatively complex information. In this regard, they endow a media outlet with personality and represent a strategy of differ-

17. www.washingtonpost.com/wp-dyn/nation/specials; http://topics.nytimes.com/topics/reference/timestopics/index.html; www.usatoday.com/news/special-reports.htm; www.cnn.com/specials; www.msnbc.msn.com/id/3032475; www.guardian.co.uk/Documentaries; www.corriere.it/Speciali; www.elmundo.es/especiales; www.elpais.com/afondo.

entiation aimed at protecting it from competition from other web publications, as well as from print, radio, and television.

Everything in these special reports fosters their differentiation from other products, starting with their design. This design aims to be original, well-executed, and attractive, to the point that these prototypes often provide a good sample set for topics related to journalistic design on the web. It might be said that no special report is quite like any other: Each one has its own home page and a navigation design adapted to its content, although it may show characteristic traits or details that become evident through repetition and that depend on each team's or media outlet's house style. The professionals who develop products of this kind are very aware of the possibilities of innovation in design and content, on the margins of the base models available for developing some special reports. In addition to their formal and design creativity, these reports are distinguished by the type of information handling they promote: provision of documents, update capabilities, and memory. Equally, they respond to novel tendencies and styles, such as audiovisualization and Journalism 2.0.

Special reports exemplify the two patterns of multimediality described in an earlier section (aggregation and integration), such that a given example can be characterized according to its greater or lesser degree of hypertextuality or multimediality, making it more of an "online story package" or a "multimedia package." The former term describes microsites or web pages bringing together a large number of links with multiple forms of relational access to independent items of content (news items, videos, audios, photo galleries, infographics, and maps); the latter refers to a product that integrates and articulates content in different media within a single interface (McAdams 2005; Kawamoto 2003).

Special reports using the first modality demonstrate one of the most complex and advanced applications of the technique of hypertextual writing for the purpose of developing an informational story. The links they employ have standardized uses, and a typology of them can be established on the basis of a variety of criteria: destination (internal and external), mode of exploration (embedded links or sidebar links), discursive purpose (development of the topic, provision of antecedents and context, documentary support, updating, access to user services), and format of the linked content (text, image, video, audio, animated file) (Larrondo 2007).

These links order micro- and macronavigation within the special website and create relationships among the contents on the basis of different types of correspondence. The discursive function they carry out can be assimilated to that of the journalistic genre of interpretative and in-depth reporting. Internal links that connect the pages of these special websites with web pages of the parent media outlets indicate that products of this kind are nourished to a large extent by content already published by those outlets previously, generally news items, although they may be conceived *ex novo*.

Planning all the material in accordance with a logical structure or navigational scheme is the most complicated task in the entire process of developing special websites. Not coincidentally, there should exist an equilibrium among user accessibility (all content on the special website should be accessible through links at any time and from different locations), informational density, and informational coherence among the interconnected parts. The documentary task of feeding the special website with content also becomes highly prominent, especially when the topic is one that remains newsworthy for an extended period.

The navigational schemes promoted by these products correspond to the hypertextual mentality that this medium demands from journalists. At the same time, the growing presence of multimedia content on these microsites has forced journalists to update their skills in order to be able to make thoughtful use of these resources, discerning when they are needed in a particular case. Journalists now have to be always on alert for the possibility of linking to an animated graphic, an audio recording, a slideshow, or an interactive multimedia product. They also need to provide spaces dedicated to the reader's direct intervention.

Although hypertextuality prevents us from speaking of a rigid structure, special websites demonstrate some conventions. Thus, they show a superstructure of mixed type, as can be deduced from an outline of their formal organization (figure 9.1). This superstructure is composed of a variety of microstructures that extend in branches or sequences of nodes, attaining significant vertical and horizontal depth. In addition, these structures interconnect with one another, enabling access between practically all the website's contents (net structure).

The other variant of the special report stands out for its specifically multimedia character. On the Internet, where the market is the same and competitors have the same tools at their disposal, some Spanish-language

Figure 9.1. Web special (Elpais.com, 2007)

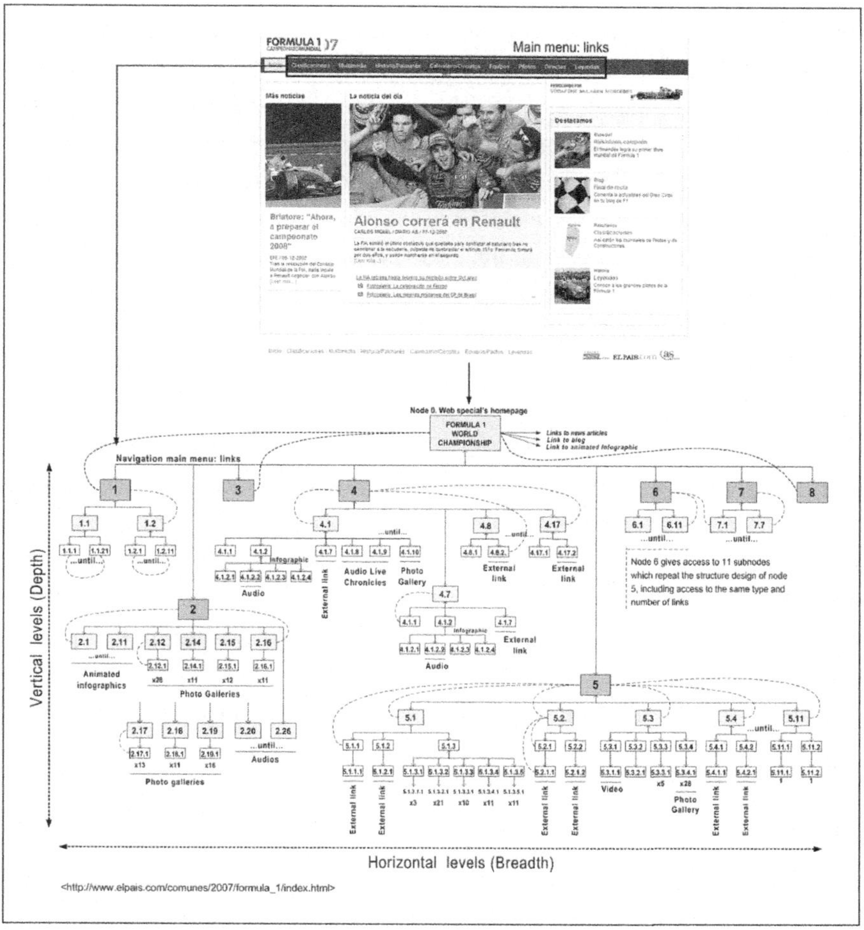

Source: Larrondo (2007).

online media, such as the Argentine *Clarín*,[18] have pioneered the development of products of this kind and earned a variety of awards for doing so.

Gustavo Sierra,[19] a journalist on the *Clarín.com* special-reports team, explains that work of this kind is based on a "constant search" for new storytelling options using multimedia, with the consequence that these projects are proposed from the perspective of innovation and always try to answer

18. http://edant.clarin.com/diario/especiales/index.html
19. Interview published in *Alternotv.com* (2009).

the question, "And what can we do now that we haven't done already?" According to Javier Elliot,[20] in charge of the section, this progressive evolution has led the team to understand that a trend as novel as multimedia narration cannot remain aloof from other equally advanced styles, such as public participation and Web 2.0. Thus, multimedia special reports are continually implementing more upload platforms so that users can make their own contributions, as well as links to social-networking sites. In the process of planning a multimedia special report, it is fundamental to choose the topic well, think about it visually, and carry out extensive pre-production and camera work. Afterward, there remains post-production and the task of assembling the entire story. Video is a fundamental resource and is combined with photography in order to take a longer look at specific scenes.

In conclusion, the need to create something new and to launch trend-setting products is especially evident in the content sphere, where new forms of creativity (new writing and new forms of presenting information) and innovation (new content models) have been fostered by online journalism and its particular language. Although it is difficult to affirm that these adapted narratives have reached their maximum development and are definitive, they do entail a considerable advance in comparison to the traditional storytelling of earlier media and offer indications of the keys to a future online journalistic language.

References

Dailey, Larry, Lori Demo, and Mary Spillman. 2005 (August). The convergence continuum: A model for studying collaboration between media newsrooms. *Atlantic Journal of Communication* 13 (3): 150–168.

Deuze, Mark. 2001 (October).Online journalism: Modelling the first generation of news media on the World Wide Web. *First Monday* 6 (10). www.firstmonday.org.

——. What is multimedia journalism? 2004 (May). *Journalism Studies* V (2): 139–152.

Dube, Jonathan. 2000 (10 July). Online storytelling forms. *Cyberjournalist.net*. www.cyberjournalist.net/news/000117.php.

20. Interview published in *Alternotv.com* (2009).

Kawamoto, Kevin. 2003. *Digital journalism: Emerging media and the changing horizons of journalism.* Oxford: Rowman & Littlefield.

Larrondo, Ainara. 2007. El reportaje hypermedia: Análisis del género en los especiales de Edmundo.es, Elpais.com y Lavanguardia.es. Ph.D. diss. Universidad del País Vasco/Euskal Herriko Unibertsitatea (University of the Basque Country).

López García, Xosé, and Xosé Pereira, eds. 2010. *Convergencia digital: Reconfiguración de los medios de comunicación en España.* Santiago de Compostela, Spain: Universidad de Santiago de Compostela.

McAdams, Mindy. 2005. *Flash journalism: How to create multimedia news packages.* Oxford: Focal Press.

Nadal, Javier, and Carlos Román. 2008 (October–December). Creatividad e innovación en la cultura digital. *Telos* 77: 48–49.

Paul, Nora, and Christina Fiebich. 2004. Elements of digital storytelling: A taxonomy of terms—and a lot of questions. In *Towards new media paradigms,* edited by Ramon Salaverría and María Rosario Sádaba. Pamplona, Spain: Eunate.

Pavlik, John. 2001. *Journalism and new media.* New York: Columbia University Press.

Salaverría, Ramón. 2005 (February). An immature medium: Strengths and weaknesses of online newspapers on September 11. *Gazette: The International Journal for Communication Studies* 67: 69–86.

Stone, Martha. 2002. Multimedia integration is here to stay. *Online Newspapers and Multimedia Newsrooms* 4: 1–4.

Thalhimer, Mark. 1994 (Winter). High-tech news or just shovelware. *Media Studies Journal* 8: 40–51.

10

Social Networks in the British Elections: Minimalist Efficiency

Iker Merodio and Estefanía Jiménez

Translated by Lauren DeAre

The British election campaign was held in April and May 2010, and was characterized by several relevant media issues, including the televised debates held between the top three candidates (in alphabetical order: Gordon Brown, David Cameron, and Nick Clegg) and the growing use of the Internet as a space to disseminate messages and strengthen the candidates' platforms.

The results of the May 6 elections in the United Kingdom of Great Britain and Northern Ireland produced a coalition government between Conservatives (led by Prime Minister David Cameron) and Liberal Democrats (led by Nick Clegg). This represented a break from certain tendencies of the previous legislatures: the consecutive victories of the Labour Party, which held power for thirteen years with three consecutive terms and absolute majority governments. In fact, in one way or another, these trends have been ongoing in the British Parliament since 1974.

The inclusion of Web 2.0 tools in the campaign was an important piece in the development of this election. Changes included televised debates, something never before seen, and a debate format that included three candidates, not just the two with the most House of Commons representation. The result of the election gave 306 seats to the Conservative Party, 258 to the Labour Party, and 57 to the Liberal Democrat Party.

In this chapter, we will examine the importance of the Internet during the electoral campaign. Robert Colville (2008) has already analyzed Nick Clegg's blog activity during his campaign to lead the Liberal Democrat Party. Therefore, in this chapter, we will present the issue and the results of observations made of the official websites of the British Conservative Party, Labour Party, and Liberal Democrat Party.

Our objective is to carry out an in-depth analysis of how these sites incorporated social networks and tools known as Web 2.0, and also take a closer look at what importance they were given. We'll also examine the changes they underwent and the number of users they had acquired as of May 5, 2010, the night before the election.

Research Objectives

Our primary objective is (O.1) to observe the value that political parties gave to social networks and similar Internet sites. This value was determined by *network placement on official websites* during the election campaign, based on the understanding that the sites were integrated into the party's communication strategies. First, (O.2) we will *identify* the social networks and similar sites where each group had a presence. That information will allow us to (O.3) *observe the differences* between the parties or, when the parties used the same tools, if they directed them at similar *targets*. This study also seeks to (O.4) measure the party's following by looking at the *number of users* that accessed the sites. The evaluation conducted will be strictly quantitative, without any intention of describing the typology of the members' network participation. Finally, we'll undertake an approximate (O.5) *analysis of the types of response and actions on the profiles* of the politicians, parties, or the so-called *community manager*.

Overview

The Internet played an important role in the long electoral campaign that landed Barack Obama in the office of president of the United States in 2008. Most of the money donated to his campaign was given through the Internet,[1] and his communication strategy was greatly facilitated by new media. The Internet advanced Obama's campaign to such an extent that the campaign has become a communications milestone.

1. See www.resourcenation.com/blog/how-obama-used-social-networking-to-set-fund-raising-records.

However, Obama was not the first or the only candidate to use social networks as communication tools for reaching potential voters (Fernández 2010). This trend is part of a general movement that includes an increased use of concepts such as interactivity and accessibility of information facilitated by websites, social networks, and user-generated content, giving the impression of greater citizen participation in democratic processes (Schultz 2000). Political parties use these environments as a way to get closer to voters and foster a feeling of community (Gould 2000). According to Eva Johanna Schweitzer (2008), the Internet allows candidates to create interactive campaigns that become platforms for debate about public policy.

Others (Anstead and Chadwick 2008) have analyzed the differences between the political system in the United States and the much more centralized British system. British centralization implies that political parties have greater power over candidate selection than in the United States, and that politicians therefore have less incentive to create a social base that supports their candidacy.

Web 2.0 tools are now fundamental parts of the *new way to do politics.* Analysts have been concerned about an obvious increase in the distance between citizens and electoral processes: In Great Britain, electoral participation was 84 percent in 1950 and fell to 59 percent in 2001 (Hall 2009). Social networks have become a way to bring political discourse closer to the electorate. Political communication supported by Web 2.0 tools has notably advanced and is now considered absolutely necessary in every campaign.[2]

Blogs and social networks are social structures that facilitate a bidirectional connection between members and are based on easy information exchange. As previously defined (Merodio 2009), all Web 2.0 sites share several defining characteristics:

1. They provide users with sites (with their own URL) that are visible to all Internet users.
2. Their design and maintenance do not require programming knowledge.
3. They are easily recognizable environments.

---

2. For information about political party participation in social networks during the election campaign period prior to the 2009 Basque parliamentary or autonomous elections, see Merodio (2009).

4. Updating them is based on systematized tools.
5. They're free, or at least the core services that allow basic use are free.
6. They allow users to establish limited or permanent relationships with other users within the network or in other networks.

According to danah boyd and Nicole Ellison (2008), a social network allows individuals to create a public or semi-public profile within a limited system, define a list of users with whom to share a connection, and view or review that list of connections and those lists made by others within the system. In this way, social network users have access to their own virtual space where they can create individualized content, show it to others, and use connections to expand their list of contacts. Internet social networks began in the 1990s, but the true emergence of the phenomena did not occur until the 2003 launch of MySpace and Hi5. Today, the amount of worldwide participation in networks is staggering. Individuals (and sometimes companies) who are employed as a part of new marketing platforms are now using networks.

Facebook, Twitter, and YouTube were the most popular Web 2.0 sites during the British campaign.

Facebook

Created in 2004 as a system exclusively designed for use in communication between university students, Facebook (FB) was opened to the general public in 2006. In July 2010, the network had five hundred million users worldwide. One month after the British elections were held, there were twenty-seven million user profiles from the United Kingdom listed on Facebook, making it the second most popular Internet site in the country—second only to the search engine Google. In addition, network use statistics indicated that in the three months prior, Facebook-directed traffic had increased by more than 10 percent.[3]

Twitter: Microblogging

Launched in 2006, Twitter allows its users to send and receive short texts. Messages are displayed on the user's profile page and are immediately sent to all users who have chosen to receive the messages. Users can link messages by theme and resend messages they want to share with their com-

3. See www.clickymedia.co.uk/2010/06/uk-facebook-statistics-for-june-2010/

munity of followers. In June 2010, Twitter had an estimated 190 million visitors monthly and its registered users posted sixty-five million "tweets" daily.[4] The figures attributed to Twitter by analysts show an incredible number of users and high levels of interest.[5] Twitter is one of the social networks that celebrities prefer to promote themselves in the public sphere, and it has helped launch virtual communities around specific issues. A large number of Twitter users employ the network for professional purposes, both as a source of information and as a communicative platform for publishing messages.

YouTube

Less than a year after its December 2005 launch, YouTube was acquired by Google for 1.65 billion dollars, a testimony to the site's capabilities. Its popularity has increased because of how easy it is to upload and share personal videos, clips, parts of TV shows, movies, and video projects. In addition to a large number of users (two billion videos viewed daily in May 2010[6]) YouTube has had an impact on popular culture and is a platform for the globalization of cultural phenomena.

Eleanor Hall (2008) considers the presence of current politicians on YouTube to be mandatory, just as in the previous century, a presence on television was mandatory.[7] According to Hall, the British political system is currently much more directed toward the Internet than when the last elections were held. In fact, in 2005, only 28 percent of British users received election information through the Internet, and neither the parties nor the majority of the members of Parliament took advantage of the possibilities available through Web 2.0 tools. By the middle of 2009, however, all the political parties in the United Kingdom had a presence on FB, MySpace, Bebo, and Twitter. The Conservative Party was the only party without a YouTube channel, but it did have a "Webcameron." The leaders of the Liberal Democrat Party and the British National Party also had personal FB profiles, and all parties had incorporated some type

4. http://techcrunch.com/2010/06/08/twitter-190-million-users/

5. http://blogs.elpais.com/estrategia-digital/2010/02/14-millones-de-usuarios-de-twitter-hacen-50-millones-de-tweets-al-d%C3%ADa-.html

6. www.google.com/hostednews/afp/article/ALeqM5jK4sI9GfUTCKAkVGhDzpJ1AC-Zm9Q

7. Used to broadcast messages that included Tony Blair expressing congratulations to Nicolas Sarkozy when the latter was elected president of the French Republic in 2007.

of Web 2.0 tool into their sites to invite visitors to make donations or suggestions.

In the same way that television and radio transformed how politicians acted in the past, the Internet will have a profound effect on the language and conduct of politicians and policies (Colvile 2008) in the future. It is an inescapable phenomenon that has already begun to occur in many countries. The amount of social network activity during the British campaign is one example.

Methodology

In order to conduct our observations, we designed a methodology framework used to identify the social networks that had links on the official campaign websites, observe the value given to these networks by the organizations, observe the similarities between them, measure user response by looking at the number of followers and observe the actions of the candidates and the parties during the campaign.

The websites analyzed were either those of the organization or the person who led the campaign. We also examined Web 2.0 sites that had a link to the candidates's main website, and included followers. Types of links were listed in order of relevance from least to most relevant: the social network identifying tag, a list of followers where some of them were identifiable, and a dynamic site with constant activity and updates.

Link placement indicated the relevance that the parties gave to the social networks: Links that appeared in the website banner or in the side column were not the same as those that could only be found by scrolling down two pages.[8]

A quantitative analysis was performed on the last day of the election campaign, and the qualitative analysis was carried out during the weeks prior to election day. These results are presented by candidate.

Gordon Brown

In general terms, Brown's campaign was the most institutional, not surprising for a candidate seeking reelection. During the campaign, the Labour Party itself dominated the network activity of the then prime minister. The general campaign strategy, which was also directed to the Internet,

8. Scrolling means vertical navigation, rolling the mouse wheel, or using the side navigation bar, allowing continuous page viewing.

presented Brown as the best choice during a crisis because of his experience in government. Of the candidates, Brown was the oldest and had the least amount of contact with the outlets. There were many fake accounts that existed on the social networks for the purpose of making fun of him, especially on Twitter. Sarah Brown, the wife of the former prime minister, also had a presence on Twitter, with more than a million followers during the period of this study.

On May 5, 2010, the day before the election, there were 37,342 users receiving updates on the Labour Party FB page; 16,252 followers of the party's Twitter feeds that had 2,090 updates; and the party's YouTube channel had 3,334 subscribers and 1,492,353 views. In all instances, the Labour Party managed the profiles and kept them focused on the group, without leaving any room for a personalized campaign from the candidate.

Efforts were directed to the most important social networks (FB and Twitter) and the largest video network (YouTube). The addition of a Flickr gallery, in which the campaign team uploaded 138 photographs between April 6 and May 6, was the exception. All images were high quality with all rights reserved, and downloading them was not allowed.

In general, the main website was straightforward for users. Deep red was combined with a light gray background, and the image of the candidate that appeared on the homepage and accompanying news was the one most regularly used. The layout was similar to a blog, with a main body in which informational pieces were organized, a banner across the top, and another banner on the right side. Links to FB and Twitter and a link for subscribing to updates appeared as buttons. Both links could be seen again as blocks in the right column with one scroll down, followed by the latest updates to Sarah Brown's Twitter account. Links to videos appeared last, in a block across the bottom, at the end of the page.

Brown's party placed all information compiled from official instruments and spokespeople on the FB network and re-sent it to the more than thirty-seven thousand recipient users. News and updates to the YouTube channel or main website refocused the central activity into the FB network. The page was programmed so that users could see only updates from administrators, not contributions from other participants. Because of this, although the "wall" was open, it was not accessible at a glance. On the other hand, Twitter was used as a way to send headlines and resend messages from other users ("retweets") or answers to them ("replies"). Sarah Brown's activity on Twitter is worth mentioning again because of the way it capitalized on the "conversation" between the candidate and the users. The prime minister's wife multiplied the number of official party account followers almost by ten, outweighing the activity of the other candidates.

David Cameron

The communications strategy that David Cameron's team undertook differed substantially from the one previously described. The Tories built their campaign around an image of change personified by Cameron himself. Therefore, both the website and the YouTube channel incorporated the candidate's name, even in the URL, and his image was more visible than the image of the party. Cameron's team used only FB and YouTube.

On the FB page, the draw was Cameron himself, who headlined the page with his first and last names and a photograph (for Brown, even the profile image was related to the party and not to the candidate). Although it looked different, the Conservative Party used its site in the same way that the Labour Party did: as a place for links and update activity. On the night before the elections, there were 36,011 users who wanted to receive updates from the site and, as of that day, the YouTube channel videos had been viewed 3,082,656 times. The candidate also had a Flickr photo gallery, but used it exactly the opposite of how Brown did. For the Labourists, Flickr was the only personalized site employed. For the Tories, it was the only site used that featured the party, not Cameron himself. The site held more than five hundred images with open licensing that were available for download.

Cameron's campaign website was also different from Brown's. The Tories offered a site that was candidate-specific and featured Cameron's biography in the central block. The left side was reserved for the party, and the right side held news and links. The color scheme of the site, blue on a white background, combined with the main photograph, managed to hide the social networks and Web 2.0 environments. The only link to FB was very discreet: an overly simple "Find David on Facebook," next to a small logo. Access to the YouTube channel was complicated because the videos were embedded into the website itself and access to Flickr didn't even appear.

As mentioned, FB became a receptacle for videos, photographs, and news sent by administrators. The site was managed for control, and user contributions were not accessible. On YouTube and Flickr, sites dedicated to images, Cameron shared the lead role with the other members of his party.

Nick Clegg

With the worst predictions and the most modest results, the Liberal Democrat candidate revolutionized the campaign both inside and outside the social networks. His appearances on television increased the media's interest in him on other outlets and channels, and on the Internet. His party worked the hardest to have a presence on the Internet and in social networks, combining new tools with more traditional media. Although Clegg appeared to be the candidate with the most high-tech profile, the image of the party was combined with the image of its leader on the various social network websites. The Liberal Democrat campaign team employed the most methods to promote the candidate and party presence on different Internet sites. The creation of a virtual card "Card.ly" that displayed links at a glance, showed the team's intentions to create a candidate that was different from the other two within the social networks.

From the card, users could access the candidate's FB page (82,273 users), Twitter account (32,939), LinkedIn professional profile, Picasa photo gallery (not updated since February 2010), a video-blog site called 12 Seconds (not updated during the campaign), and YouTube videos (the only site that was party-related and not candidate-related; with 2,860 subscribers). Along with this collection of sites, there was also a Flickr photography gallery and, of course, the personal website of the campaign. Liberal Democrats also had sites that pertained to the party, not to the candidate. It's noteworthy that Clegg's campaign team neglected websites that were directly related to images (Flickr, Picasa, and 12 Seconds), with the exception of YouTube, in which all the candidate's telegenic talent and the party's presence were concentrated.

During the campaign, the Liberal Democrats chose to maintain two websites: one for the party and another for the candidate, as they also did on FB and Twitter. Both had a very similar layout. Clegg's was the more basic one. It was fully visible with just two scrolls down and featured three columns.

Clegg's campaign website layout drew the viewer's attention strongly toward the center block, where the candidate offered his personal guarantee for the development of liberal and democratic policies. The links to Clegg's FB site and Twitter and the party's YouTube channel were located at the bottom-right side of the page, overly discreet for a candidate with a presence in multiple networks. The site was focused on offering the agenda, using divided areas that covered the entire page and the previously mentioned personal guarantee in the center. Links were more noticeable on the party's website, and were located on the top of the right column and visible without having to scroll down. The links all connected to the FB page, Twitter, and YouTube channel that belonged to the party.

Comparison

Based on the number of users that joined his FB page (82,273), Nick Clegg held a large advantage in networks and numbers when compared to Brown (37,342, although it was the party's page and not the candidate's) and, trailing him by just a bit, Cameron (36,011). On Twitter, however, the Conservatives held the advantage (36,310 followers) over Liberal Democrats (19,353) and Labourists (16,252), although the latter two parties divided their efforts on the network with the accounts of Nick Clegg and Sarah Brown. On YouTube, followers are not as important. What counts is if content is systematically republished through links and by reposting the clip itself on the other two networks.

The strategies of the three parties and the way they approached their websites were very similar. Gordon Brown was the only candidate who didn't create a personal site in addition to the party's. The other two candidates created sites that were very similar to the parent site: very concise and with few links. The Labourists gave the most importance to network links, providing differentiated blocks in the right column and across the bottom. Liberal Democrats used the network logos on their website and they were well-placed. On the candidate's page, the icons were smaller and were given worse positioning. Finally, the Conservatives did not put links to the sites on their website or on the candidate's site, only a small link to FB was visible.

Despite differences in the different tools implemented, how links were used, and the distribution of candidate presence and how it was organized, the actions taken by the three parties were similar. The practices and objectives that the groups undertook were choosing the foremost networks; abandoning secondary ones; offering opportunities to repost content; and, above all, broadcasting campaign videos.

Conclusions

The major British political parties decided to use the social networks during the campaign, but they did it in a minimalist way, seeking the networks that had the most public potential and concentrating their efforts. Although no clear pattern was established regarding whether the candidate or the party led the website presence, it appears clear that in no case was the person the sole attraction, and that the party was always present (in some cases, such as Gordon Brown's campaign, more so than the candidate himself).

The three parties directed their campaigns in completely different ways. The permanent antagonism between Labourists and Conservatives stood out. Even though they each used very few sites, the ways the parties used them were diametrically opposed. Liberal Democrats, having started out with fewer options available, employed the most use of the networks.

Facebook, Twitter, and YouTube were the sites most utilized and linked to the official campaign websites and were used as a way to continually rebroadcast information and campaign video clips. The parties did not take risks in their use of FB. On their pages, by default, the user could always see what the site managers published, not what other visitors published. This implies what little confidence the campaign teams had in the tool. Along those lines, blogs tended to disappear, replaced by traditional websites that could represent the parties' platform and policies. FB and Twitter allow sending short messages and links, and the way all of the candidates chose to use them showed that the parties preferred "headlines."

References

Anstead, Nick, and Andrew Chadwick. 2008. Parties, election campaigning and the internet: Toward a comparative institutional approach. In *The Routledge handbook of internet politics*, edited by A. Chadwick and P. N. Howard. Oxford: Routledge.

boyd, danah, and Nicole Ellison. 2008. Social network sites: Definition, history and scholarship. *Journal of Computer-Mediated Communication* 13 (1): 210–230.

Colvile, Robert. 2008. *Politics, policy and the internet*. London: Centre for Policy Studies.

Fernández, Carmen Beatriz. 2010. Ciberpolitiqueando. *Campaigns & Elections* 8 (August): 14–16.

Gould, Philip. 2000. Parliament and the electronic media. *Royal Society of Arts Journal* 1: 70–72.

Hall, Eleanor. 2009. *Politics in the YouTube age: Transforming the political and media culture?* Oxford: Reuters Institute for the Study of Journalism.

Merodio, Iker. 2009. Euskadi 2009: La prueba cero de las campañas digitales después del fenómeno Obama. *Hermes* 30: 50–58.

Schultz, Tanjev. 2000. Mass media and the concept of interactivity: An exploratory study of online forums and reader email. *Media, Culture and Society* 22, no. 2: 205–21.

Schweitzer, Eva Johanna. 2008. Innovation or normalization in e-campaigning? *European Journal of Communication*, 23 (4): 449–70.

11

New Proposals for the Creation of an Informative Agenda: The Case of Menéame

EVA FERRERAS

Translated by Jennifer Martin

New tools emerged from the Web 2.0 philosophy. The term's origin was accepted as coming from a brainstorming session between the O'Reilly Media and MediaLive International teams in the middle of 2004. After its creation, it spread rapidly through the Internet without a clear definition existing for the term. For this reason, in 2005, Tim O'Reilly theoretically explains it in the article "What Is Web 2.0: Design Patterns and Business Models for the Next Generation of Software" (2005).

O'Reilly points out that "like many important concepts, Web 2.0 doesn't have a hard boundary but, rather, a gravitational core. You can visualize Web 2.0 as a set of principles and practices that tie together a veritable solar system of sites that demonstrate some or all of those principles, at a varying distance from that core" (2005, 2). He cites seven constituent principles: the web as platform, harnessing collective intelligence, database management as core competency, the end of the software release cycle, lightweight programming models, software that is not limited to a single device, and rich user experiences.

On the other hand, various interpretations of the term coined by O'Reilly exist. Tim Berners-Lee, creator of the World Wide Web, and Paul Anderson (2007) state that Web 2.0 should be seen as an evolution or as a consequence of the mere implementation of all Web 1.0 services.

Other authors think that it was a term devoid of meaning that only corresponded to a marketing strategy. Andrew Keen (2007) and Nicholas Carr (2005) think that Web 2.0 promotes collective amateurism, increasing noise on the Internet.

Despite the fact that a fully valid definition does not exist, beyond the numerous interpretations it is possible to sum up the concept in three fundamental points, which according to Ismael Nafría (2007, 14) are the following:

1. Second phase of the Internet: Web 2.0 is the second stage of projects and business on the Internet, after what was called the "dotcom bubble" in 2000.
2. Web as platform: a new way of offering services over the Internet, thanks to the addition and combination of several technologies that allow the Net to be used as a platform.
3. The user is king: from being a mere consumer to becoming the creator of content and services.

In the field of journalism, the expansion of new technologies has brought about a substantial change in the relationship that readers/users have traditionally held with the media. The hatching of social media and the emergence of new applications have created an opening for citizen participation in the media. Thus, concepts such as "participatory journalism" have proved defining to journalistic activity of the twenty-first century.

The well-known report "We media," by Shayne Bowman and Chris Willis (2003), examines the idea of audience participation in compiling information versus large media companies. Here, participatory journalism is defined as "The act of a citizen, or group of citizens, playing an active role in the process of collecting, reporting, analyzing, and disseminating news and information. The intent of this participation is to provide independent, reliable, accurate, wide-ranging, and relevant information that a democracy requires" (Bowman and Willis 2003, 9).

Moreover, Dan Gillmor has coined the nomenclature "Journalism 3.0," a third version of cyber journalism that refers to the socialization of journalistic information obtained through digital tools. "Journalism 1.0" transferred the contents of traditional media to the Net; "Journalism 2.0" was Internet content creation, paying attention to the characteristics of hypertextuality, interactivity, and multimedia. Juan Varela (2005), journalist, blogger, and successor of said denomination in Spain, also understands the era of Journalism 3.0 as one of participatory journalism.

In 2006, the Spanish digital press began a process in which it increased and diversified the participatory applications and tools that were provided to its readers: sections of blogs, RSS (Really Simple Syndication) channels, applications that allowed users to send their own news, photos, video inclusion, and so on. It should be noted that one of the first applications that the media installed in this process was the news rating icons for websites such as Menéame or Digg.

Even though it is true that the expansion of daily digital newspapers occurred against an admittedly optimistic backdrop, the global economic crisis that arose during the last few months of 2008 drastically impacted the media business.

Print media was the hardest hit during the crisis, and there was also an obvious impact on cyber media, especially those with printed versions. The growth seen in previous years of digital editions slowed down and companies failed to sustain a profit to compensate their global losses (Guallar 2009, 88).

Social News Promotion and Communication Media

"Social news promotion" websites can be defined as collaborative applications in which the users propose and vote on content that was previously published in blogs and other cyber media forms, whether they were news items, videos, or even podcasts. These contents are hierarchically structured and ordered on the home page according to the number of votes cast by the users themselves. Any registered user can send in an excerpt from a news article published on another site, and other users, registered or not (a certain percentage of anonymous votes are allowed), can vote either positively or negatively on the item according to their interests. When news items receive a sufficient amount of votes, weighted by a factor associated with each registered user known as *karma*, they are promoted to the home page, obtained as a combination of the collective interests.

According to Kristine Lerman (2007, 30), these applications are characterized by the following elements:

- The users create information.
- The users comment on the information with key words or labels.
- The users evaluate the content, either passively or actively.

- The users create social networks, designating other users with related interests as friends or contacts.

The same users are responsible for collecting the contents, posting them on the web, and then rating them, usually by means of votes. Those that obtain the highest ratings go on the homepage, which means, at least a priori, that they reflect the interests of the community. That is, everything is focused on user participation, which in turn is rewarded with *karma,*[1] and on the use of collective intelligence.

The antecedents of these systems are found in the English-speaking market, with Slashdot and Digg as main benchmarks. The success of Digg, which was developed in 2004, prompted the creation of a similar application for the Spanish-speaking market: Menéame.

There is no mediation on these websites: it is not a professional who chooses the contents, but the community. Therefore, the user could be the editor who posts the content that he wants, assesses the articles by means of voting, and also comments on them.

For their part, Enrique Orduña-Malea and Juan Antonio Ontalva-Rupiérez (2008, 2) proposed the following stages in the evolution of "social news compilation services":

- Automatic structuring, manual compilation, and controlled posting: corresponding to the phenomenon known as recirculation and content aggregation, and should not be confused with content aggregation services made by intermediaries.
- Structuring and manual compilation, controlled posting: social news compilers: This stage distinguishes itself from the previous one in that the metadata describing the news is no longer created automatically at the source (in the medium), but by the users themselves at the destination. Posting is controlled because there is an editorial staff that approves or rejects posts. The pioneering service was Slashdot, created in 1997, and the website Barrapunto was created in 2000 in Spain.
- Structuring, compilation, and social posting: social media management tools. This involves the complete alignment of journalistic communication models with Web 2.0.

---

1. A quantitative measure that establishes the value or rank of an individual/user in the community. *Karma* is usually obtained in communities of this type by means of contributions, quality of comments, and content voting. Value is obtained through a mathematical algorithm.

One of the first elements that Spanish newspapers incorporated into their digital editions during this adaptation process to the new Internet phase was a social media management tool. This implementation was made by placing a plug-in, also called the site's "social button" after each news item. Figure 11.1 shows the icon that identifies Menéame.

Figure 11.1. Menéame icon

Source: www.meneame.net.

Menéame is a project of Ricardo Galli, a professor in the Computer Science Department at the University of the Balearic Islands, and Benjamí Villoslada. It was developed using PHP, MySQL, and AJAX; made public in December 2005; and released as Free Software licensed under Affero GPL.

Galli acknowledged that the blueprint for developing Menéame was Digg, but he introduced some changes; he released the code, included a trackback service in order to facilitate conversation, and allowed anonymous voting by non-registered users (Galli 2005).

Regarding Menéame:[2] as of September 2009 it had 2.5 million unique users according to Google analytics; 121,000 of those users held active accounts; 690 media were sent and 6,236 comments were made daily, as well as 60,048 votes.

During these years, the creators of Menéame were progressively making improvements in the comment system, in the algorithm structure for calculating karma and promoting news, and they even developed a version for cellular phones.

Initially, Menéame was centered around science and technology content, but by the end of 2007, the thematic meta-categories of technology,

2. Data provided by Ricardo Galli.

culture, and current events were introduced. A "leisure time" category was added later.

When a post is first sent, it goes to a "pending list," a section that is visible to all users. Any user can vote on the contents from the site's home page as well as from the pending list, but one must be registered in order to post content, comment or cast a negative vote.

Every registered user is assigned *karma*—points that are calculated by means of an algorithm and that determine the user's rank and participation. It is a value that goes from one to twenty, with six being the number that is given upon registering. The algorithm that calculates the points takes the previous four days of activity into account: votes received for posted content, positive votes awarded to other posts, negative votes, and votes on your comments.

Essentially, *karma* is a way of exercising control and is based on the user's merits as described above; therefore it is considered to be a meritocratic system. At the same time, it has been one of the most criticized characteristics of Menéame. According to David de Ugarte (2007, 45): "The logic of incentives in these types of services inevitably leads to the formation of relatively stable participatory oligarchies. The most radical examples would be Web 2.0 services, such as Menéame or participatory media. In these, a small, stable group will create content and a larger mass of passive users will consume and disseminate it."

It is also important to be aware of the audience data related to Menéame. According to Google Analytics, in September 2009, the number of absolute unique users was 2.5 million.

We can also look at the data from the Nielsen Netview panel from February 2008 to February 2009 (table 11.1).

Table 11.1. Menéame (February 2008–February 2009)

Report: Trend Report (Internet Applications Included) 13 Months Period: Month of February, 2009 Panel Type: Home and Work Country: Spain

Trending : Unique Audience (000)

Site	Level	feb-08	mar-08	abr-08	may-08	jun-08	jul-08	ago-08	sep-08	oct-08	nov-08	dic-08	ene-09	feb-09
Menéame	Parent	506	515	456	481	442	535	473	608	742	815	592	781	828

Source: Nielsen Netview.

On the other hand, the most recent data from April 2010, also from Nielsen Netview, was the following:

- Unique users: 704,000
- Page visits: 16,808,000

Digital Press and Menéame

As this system acquired prominence, the Spanish digital media began to add plug-ins that made it easier for their readers to send posts to Menéame. The most-represented media on Menéame are digital newspapers. I demonstrated this clearly in my dissertation, which I defended in April 2010. I carried out research between February 22 and March 10, during the electoral period coinciding with the Spanish general elections in 2008. We can see that the first six forms of media on the chart are digital newspapers (figure 11.2).

Figure 11.2. Menéame media rankings

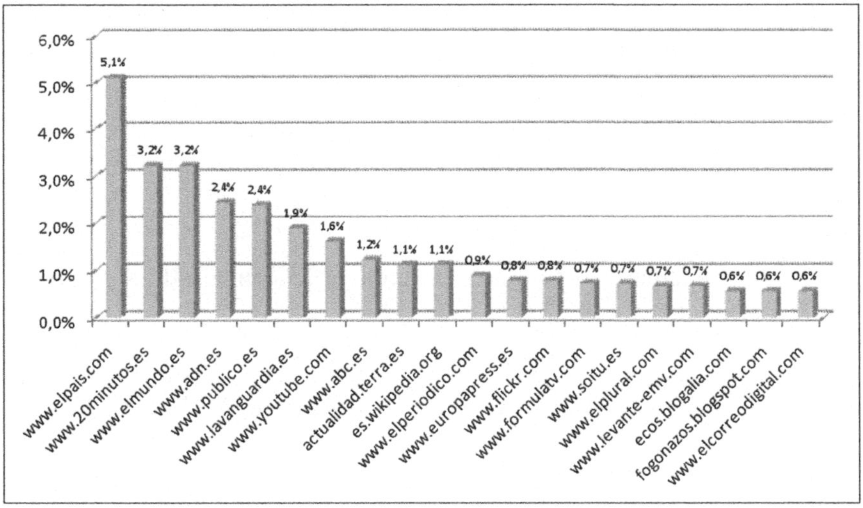

Source: Ferreras (2010), 290.

The data presented may seem far removed from the present time, and even somewhat inflated because it was collected during an electoral period in which the amount of journalistic information was considerably higher. Nevertheless, if we focus on current data from the site itself, we see that media appear every day, newspapers above all, in the ranking of most-voted-for sites. For example, on July 25 at around 11 a.m., the listing of the most-voted-for sites that appeared on the main page of Menéame included four newspapers: *El Mundo, El País, Público,* and *El Periódico de Catalunya.* Other media sites also appeared: the website for Radio Television Española, one for the news agency Europa Press, and one for the free newspaper from Valparaiso, Chile.

On this subject, Galli maintains: "Clearly the most prevalent media on Menéame are the largest, *El País, El Mundo, 20 Minutos* . . . and Menéame placed restrictions in this respect by not allowing you to post more than four percent of content from a single site; this was done in order to avoid a monopoly of content from certain sites" (Ferreras 2010, 449).

Returning now to the media in order to understand their reasons for incorporating social sharing icons such as the one belonging to Menéame. Those responsible for the digital editions of *El Mundo, El País, 20 Minutos,* and *ABC* acknowledge that one of the motives is the user traffic that these icons bring back to them. According to the data provided by Nielsen on Menéame's outgoing traffic, also from February 2008, seven newspapers were among the destinations that Menéame reported as having the most traffic. From these figures, and knowing the audiences both from the site and from the newspaper, it was possible to form a calculation to find out, according to the data provided by Nielsen, the percentage of visits that Menéame returns to the newspapers (Ferreras 2010, 326). This can be seen in table 11.2.

Table 11.2. Calculation of traffic reported by Menéame in three periods in February 2008

Media	Menéame audience	% visits to the media from Menéame	Total visitors to the media	% of visits attributable to Menéame
20 Minutos	505,747	16.08	1,453,309	5.6
El Pais	505,747	13.63	2,176,106	3.17
El Mundo	505,747	10	3,078,126	1.71

Source: Ferreras (2010), 326.

New Role of the User/Prosumer

The academic community has studied this new role at great length. For example, Octavio Islas-Carmona (2008, 20) contends that the user or Internet surfer "acceded to the status of *prosumer*" with the arrival of Web

2.0 services. *Prosumer* is a compound that comes from fusing together two words in English: *producer* and *consumer*.

However, several paradoxes surrounded this idea of participation. Many studies and analyses showed that participation rates were poor.

Bill Tancer is the manager of global research at Hitwise and author of *Click: What Millions of People are Doing Online and Why It Matters* (2008), in which he dedicates a section to the "participatory" web. Here, he reports on the results of user participation from three Web 2.0 sites: YouTube, Flickr, and Wikipedia. The study shows that only a small percentage posted videos, photos, or texts—that is, a few users monopolize activity while the rest simply view the pages, without contributing.

The term "lurker" is used to refer to those virtual community participants whose only activity is receptive, devoid of active contributions.

On Menéame for example, only 7.4 percent participate by posting content, 20 percent comment, and the rest vote or only read what is posted (Ferreras 2010, 349). Thus, there is an active minority, while a large part of its users are considered to be lurkers. This issue is closely linked with *karma* and the creation of elites in these communities. If *karma* is indeed the system by which user participation is ranked, and this participation is very small, it is obvious that the actively participating minority builds up more *karma*, which means that their votes and comments are "worth" more. In this way, their influence on the site is very much like a "reward" for their high level of involvement.

Implications for Journalism

Another question remains to be studied: how the use and popularization of tools like Menéame have influenced journalism. In the first place, the way information is circulated has changed; there are now other channels for distributing news, which was once the capacity of the traditional media alone. The websites studied do not practice journalism, nor are their users journalists; they are limited to posting and prioritizing the information of others. Perhaps the most important aspect is that these websites can quickly and easily create places for debate, discussion, and conversation, something that no other means of communication has been able to achieve online until now. The media have lost their control over the information and also over their distribution channels on the Internet.

It has been proven that digital newspapers have an extremely large presence on Menéame, partly due to the increase of *plug-ins*, or social sharing buttons, on their digital editions. Media outlets did not create these tools, but they "sponsor" them mainly for the traffic that they bring in, through what is known as the "Menéame effect." That appears to be reason enough for their involvement, but there is more: the need to keep up in a rapidly evolving market where readers demand those types of participatory applications from the press. In the short span of approximately four years, the digital press in Spain went from being scarcely developed to promoting user participation as an indispensable value and striving to be open to different forms: hosting blogs, allowing comments, creating applications so that users can post their information, or facilitating deliveries to content management systems and social networks.

Clearly, Menéame's homepage represents a source of alternative information to the media; 67.6 percent of its users say that they use it to read the content and to participate, while 31.5 percent are not interested in participating and only read the content. Furthermore, 69.2 percent check the homepage several times a day.

This implies that for the first time in history and through systems such as Menéame, for example, readers are likely to be informed on subjects that interest them without having to turn to mass media.

Furthermore, while there is a mandated hierarchy for deciding on content in the media, on Menéame the editorial power is in the hands of the community that filters content in a democratic and collaborative way. The role of *gatekeeper*, traditionally played by the media, has been placed into the hands of the Internet community concerned.

On several occasions, content that did not figure prominently in the media but that the public liked appeared on the homepage; proof of this is that they were the most-read items. The most-read or most-voted-for articles within the media do not always appear in the headlines or in preferential spaces. This demonstrates what has been discussed ad nauseam about the lapse between what is published and the public's interest. Maybe it is because of this that sites like Menéame work because its users feel as if they read and rank that which interests them both individually and as a community.

Conclusions

Active user participation in public discourse is not a fleeting phenomenon, and digital media will open up its sites more and more to this participation. It does not appear that rating systems are leaving the media, either;

several Spanish newspapers have recently updated their digital editions, and the social buttons were maintained. The aggregators will remain useful to the media as long as they support traffic, however slight, and while they maintain the ability to amplify content.

It is obvious that these types of collaborative tools create an alternative agenda to the media. José Luis Orihuela (2009) observes, "Reviewing Menéame's homepage on any given day or comparing the homepage of any newspaper's online edition to the reader's homepage (the most-voted for or most-viewed items also), reveals the extent to which the agenda and interests of the audience and of the media diverge. Nevertheless, alternatives must continue to be explored that effectively utilize collaborative filtering to enrich editorial decisions."

Accordingly, in the prologue to *Planeta 2.0* (Cobo and Pardo 2007, 22), Alejandro Piscitelli explains: "Neither Digg nor Menéame, nor the hundreds of Web 2.0 tools inventoried in this book, are the marvel that was promised, nor the trifle that they are pronounced to be today. They are barely primitive forms of new technocultural embryos, which once reinvented and retested, will offer a powerful combination of the old analog culture's quality (customized and individualized), with creative power and social rating, and of (collaborative) filtering from the new digital culture" (2009, 22).

However, there are still many changes necessary in order for the embryo to mature: First of all, social rating from collaborative filtering is entirely a value that these websites add on, but let's not forget that the active participation figures are not so high. Furthermore, it remains to be seen how the generations called "digital natives" will utilize these types of applications, which with all certainty will technologically improve and possibly propose different interfaces for each user, just as David de Ugarte (2007) suggests.

The low participation rate that has already been mentioned may result in the strengthening of the elites in the community if it is not properly managed. On the other hand, the abuses (spam) or misapplications that develop in these systems may jeopardize projects of this type.

For the time being, its success is undeniable and imitations have multiplied across the Internet.

References

Anderson, Paul. 2007. *What is Web 2.0? Ideas, technologies and impli-*

cations for education. www.jisc.ac.uk/whatwedo/services/services_techwatch/techwatch/tech-watch_ic_reports2005_published.aspx.

Bowman, Shayne, and Chris Willis. 2003. We media: How audiences are shaping the future of news and information. *The Media Center-American Press Institute*. www.hypergene.net/wemedia/weblog.php (accessed June 7, 2009).

Carr, Nicholas. 2005. The amorality of Web 2.0. www.roughtype.com/archives/2005/10/the_amorality_o.php

Cobo, Cristóbal, and Hugo Pardo. 2007. *Planeta Web 2.0: Inteligencia colectiva o medios fast food*. Barcelona/México DF: Grup de Recerca d'Interaccions Digitals, Universitat de Vic; Flasco México.

de Ugarte, David. 2007. *El poder de las redes*. Barcelona: Colección Planta 29.

Ferreras, Eva María. 2010. "Herramientas Web 2.0 para la promoción social de noticias en España. Estudio y análisis de Menéame, Fresqui y Negóciame durante las elecciones generales de 2008." Ph.D. diss., Universidad del País Vasco/Euskal Herriko Unibertsitatea (University of the Basque Country).

Galli, Ricardo. 2005 (8 December). Qué y porqué del Menéame. http://mnm.uib.es/gallir/posts/2005/12/08/535/

Guallar, Javier. 2008. Prensa digital en 2007. *Anuario ThinkEPI, EPI SCP*: 84–99.

Islas-Carmona, Octavio. 2008. "El prosumidor. El actor comunicativo de la sociedad de la ubicuidad." *Palabra Clave* 11, no. 1: 29–39.

Keen, Andrew. 2007. *The cult of amateur: How today's internet is killing our culture*. New York: Doubleday/Currency.

Lerman, Kristine. 2007. Social information processing in social news aggregation. *IEE E Internet Computer* 11 (6): 16–28. http://arxiv.org/PS_cache/cs/pdf/0703/0703087v2.pdf; arxiv.org/PS_cache/cs/pdf/0703/0703087v2.pdf

Lévy, Pierre. 2004. *Inteligencia colectiva por una antropología del ciberespacio*. Washington: Organización Panamericana de la Salud, Unidad de Promoción y Desarrollo de la Investigación y el Centro Latinoamericano y del Caribe de Información en Ciencias de la Salud.

Nafría, Ismael. 2007. *Web 2.0: El usuario el nuevo rey de internet*. Barcelona: Gestión.

Orduña-Malea, Enrique, and Juan Antonio Ontalba-Ruipérez. 2008.

Presencia de la prensa digital española en la web social: Análisis de Menéame. *El profesional de la información* 17 (5): 87–98.

O'Reilly, Tim. 2005. What is Web 2.0: Design patterns and business models for the next generation of software. *O'Reilly Media.* www.oreillynet.com/pub/a/oreilly/tim/news/2005/09/30/what-is-web-20.html

Orihuela, José Luis. 2009 (30 November). ¿Hacia dónde nos llevan las audiencias?"*ABC.* www.abc.es/blogs/jose-luis-orihuela/public/post/hacia-donde-nos-llevan-las-audiencias-2433.asp.

Piscitelli, Alejandro. 2009. *Nativos digitales: Dieta cognitiva, inteligencia colectiva y arquitecturas de la participación.* Buenos Aires: Ediciones Santillana S.A.

Tancer, Bill. 2008. *Click: What millions of people are doing online and why it matters.* New York: Hyperion.

Varela, Juan. 2005. Periodismo 3.0, la socialización de la información. *Telos* 65. http://sociedadinformacion.fundacion.telefonica.com/telos/articulocuaderno.asp@idarticulo=7&rev=65.htm

Index

J

K

N

O

List of Contributors

For full biographical information about the contributors, links to their projects, and more, visit www.basque.unr.edu/currentresearch/contributors

Pedro Abramo
Iñaki Alcedo Momoitio
Alejandro Arizkun Cela
Sabino Ayestarán Etxeberria
Jose M. Barrutia
Carmen Echebarria
Goio Etxebarria
Virginia Fernández
Eva Ferreras
Mikel Gómez Uranga
Estefanía Jiménez
Ainara Larrondo Ureta
Iñaki Lasagabaster Herrarte
Iker Merodio
Juan Carlos Miguel de Bustos
José Moreno Díaz
Francisco Panera Mendieta
Josu Rekalde
Arantxa Rodríguez
Itziar Urizar
Mikel Zurbano

Made in the USA
Monee, IL
06 November 2026

68002975R00114